Interviewing in Criminal Justice

Interviewing in Criminal Justice

Victims, Witnesses, Clients, and Suspects

Vivian B. Lord, PhD
Department of Criminal Justice and Criminology
University of North Carolina, Charlotte

Allen D. Cowan
Cowan's Private Investigative Service

JONES AND BARTLETT PUBLISHERS
Sudbury, Massachusetts
BOSTON   TORONTO   LONDON   SINGAPORE

World Headquarters

Jones and Bartlett Publishers
40 Tall Pine Drive
Sudbury, MA 01776
978-443-5000
info@jbpub.com
www.jbpub.com

Jones and Bartlett Publishers
Canada
6339 Ormindale Way
Mississauga, Ontario L5V 1J2
Canada

Jones and Bartlett Publishers
International
Barb House, Barb Mews
London W6 7PA
United Kingdom

Jones and Bartlett's books and products are available through most bookstores and online booksellers. To contact Jones and Bartlett Publishers directly, call 800-832-0034, fax 978-443-8000, or visit our website, www.jbpub.com.

Substantial discounts on bulk quantities of Jones and Bartlett's publications are available to corporations, professional associations, and other qualified organizations. For details and specific discount information, contact the special sales department at Jones and Bartlett via the above contact information or send an email to specialsales@jbpub.com.

Production Credits
Publisher, Higher Education: Cathleen Sether
Acquisitions Editor: Sean Connelly
Associate Editor: Megan R. Turner
Production Manager: Julie Champagne Bolduc
Associate Production Editor: Jessica Steele Newfell
Associate Marketing Manager: Jessica Cormier
Manufacturing and Inventory Control Supervisor: Amy Bacus
Composition: Glyph International
Cover Design: Kristin E. Parker
Cover Image: © Jones and Bartlett Publishers
Printing and Binding: Malloy, Inc.
Cover Printing: Malloy, Inc.

Library of Congress Cataloging-in-Publication Data
Lord, Vivian B.
    Interviewing in criminal justice : victims, witnesses, clients, and suspects / Vivian B. Lord, Allen D. Cowan.
        p. cm.
    Includes bibliographical references and index.
    ISBN 978-0-7637-6643-6 (pbk. : alk. paper)
    1. Interviewing in law enforcement. I. Cowan, Allen D. II. Title.
    HV8073.3.L67 2010
    363.25'4—dc22
                                                        2009053796
6048

Printed in the United States of America
14 13 12 11 10   10 9 8 7 6 5 4 3 2 1

Brief Contents

PART I   **BEFORE INTERVIEWING**   **1**

     Chapter 1   Foundations of Interviewing   3

     Chapter 2   Ethics Surrounding Interviewing   19

     Chapter 3   Preparing to Interview   35

PART II   **CONDUCTING THE INTERVIEW**   **61**

     Chapter 4   Beginning the Interview   63

     Chapter 5   Conducting the Interview   95

     Chapter 6   Interview Documentation   123

PART III   **SPECIAL AREAS**   **145**

     Chapter 7   Interviewing Victims: Adults and Children   147

     Chapter 8   Cultural Differences   167

     Chapter 9   Interviewing for Defense Attorneys   189

     Chapter 10   Interrogation   209

     Chapter 11   Crisis Intervention and Negotiation   237

     Chapter 12   Polygraph Use   257

     Chapter 13   Stance-Shift Analysis   273

PART IV   **INTEGRATION**   **283**

     Chapter 14   Putting It All Together   285

     References   295

Contents

PREFACE   **xv**

ACKNOWLEDGMENTS   **xix**

INTRODUCTION   **xxi**

PART I   **BEFORE INTERVIEWING**   **1**

Chapter 1   **FOUNDATIONS OF INTERVIEWING**   **3**
Basis of Communications   3
   Communication as a Process   5
Inhibitors of Effective Communications   9
Human Needs as Motivators   11
Reducing Anxiety   12
Principles of Interviewing   14
   Individualization   14
   Curiosity   14
   Flexibility   14
   Self-Fulfilling Prophecy   15
   Self-Confidence   15
   Expression of Feelings   16
   Self-Determination   16
   Confidentiality   16
Conclusion   16
Allen's World   17

Chapter 2   **ETHICS SURROUNDING INTERVIEWING**   **19**
Introduction   19
Ethical Standards   20
Validity of Confession Evidence   22
Ethical Tools   25
Confidentiality   26
Conclusion   27
Allen's World   29

Chapter 3   **PREPARING TO INTERVIEW**   **35**
Introduction   35
Interview Objectives   36
Formulating Basic Questions   38
    Words   40
    Formulating Questions That Motivate Responses   40
    Empathy   42
    Formulating Questions That Enhance
        Memory Retrieval   43
Types of Questions   48
    Open-Ended Questions   48
    Closed-Ended Questions   49
    Sequencing Types of Questions   50
    Nonthreatening Questions   51
Questions to Avoid   54
    Why Questions   54
    Questions That Lead to Confusion   55
    Leading Questions   55
Conclusion   55
Allen's World   56

PART II   **CONDUCTING THE INTERVIEW**   **61**

Chapter 4   **BEGINNING THE INTERVIEW**   **63**
Introduction   64
Interview Phases   64
    Phases of the Investigative Interview   64
    Phases of the Helping Interview   68
Establishing Rapport   69
    Mirroring   72
Listening   73
    Difference in Hearing and Listening   73
    Obstacles to Listening   74
    Attending Behavior   74
Nonverbal Communication   76
    Interviewer's Nonverbal Communication   76
    Use of Silence   77
    Interviewee's Nonverbal Clues   77
Neurolinguistics: Language of the Mind   84
    Building Rapport   85
    Use of Lateral Eye Movement   86
Physical Settings   88
    Interview Area   88
    Seating Arrangement   89
    Other Physical Setting Concerns   89

Verbal Setting   90
Conclusion   91
Allen's World   92

Chapter 5   **CONDUCTING THE INTERVIEW**   **95**
Introduction   95
During the Interview   95
   Motivating Witnesses   95
   Empathy   97
Reflecting and Accepting Content   98
Reflection of Feelings   100
Evaluating Information   104
Probing   107
Conducting Helping Interviews   111
   Resistant Clients   114
Conclusion   119
Allen's World   119

Chapter 6   **INTERVIEW DOCUMENTATION**   **123**
Introduction   123
Recording and Note Taking   124
Probe Notes   125
The Statement   126
Formal Statements   127
   Statement Format   128
   Signature and Correcting Errors   129
   Challenges to Statements   129
Interviewer's Report   131
Investigative Discourse Analysis   132
   Analysis of the Statement's Structure   133
   Mean Length of Utterance   134
   Semantic Analysis   135
Conclusion   141
Allen's World   142

PART III   **SPECIAL AREAS**   **145**

Chapter 7   **INTERVIEWING VICTIMS:**
                   ADULTS AND CHILDREN   **147**
Introduction   147
Interviewing Adult Victims   147
Interviewing Children   151
   Children's Developmental Levels   152
   Children's Cognitive Factors   153

Repressed Memory   155
Language Development   155
Social Factors   156
Building Rapport   157
Physical Setting   157
Verbal Setting   157
Conducting the Interview   157
Formulating Questions   159
Maximizing Facilitators—Minimizing Inhibitors   159
Terminating the Interview   161
Conclusion   161
Allen's World   162

Chapter 8   CULTURAL DIFFERENCES   167
Introduction   167
Dimension of Cultural Variability   168
Individualism Versus Collectivism   169
High Context and Low Context   170
Horizontal Versus Vertical Cultures   171
Masculinity/Femininity   172
Perception of Time   173
Nonverbal Communication Differences   175
Body Space   176
Facial Expressions   177
Eye Contact   178
Effective Intercultural Communication   179
Developing Rapport: A Good Place to Start   180
General Hints for Effective Intercultural
Communication   180
Conclusion   183
Allen's World   184

Chapter 9   INTERVIEWING FOR DEFENSE ATTORNEYS   189
Introduction   189
Use of Private Investigators by Criminal
Defense Attorneys   190
One District Attorney's Perception of
Private Investigators   191
The Role of the PI for Defense Attorneys   192
Domestic   193
Personal Injury   197
Criminal Defense   202
Conclusion   206

Chapter 10   **INTERROGATION**   **209**
Introduction   209
Interrogation Defined   209
Precautionary Measures   210
The Psychology of Confessions   211
    Deception   213
Legal Framework of Interrogations   216
    Fourth Amendment   216
    Fifth Amendment   216
    Sixth Amendment   218
    Fourteenth Amendment   219
    Permissible Interrogation Tactics and
        Techniques   219
    Emergency Exception to *Miranda*   221
    Youthful Suspects   221
Major Interrogation Strategies   222
    Important Principles   222
    Introductory Statements   223
Interrogation Approaches   224
    Factual Approach   224
    Indirect Approach   227
    Reid Nine Steps of Interrogation   228
Additional Interrogative Techniques   230
    Implication Questions   230
    Trap Questions   231
    Defeated Protests   231
Ending the Interrogation   232
Conclusion   233
Allen's World   234

Chapter 11   **CRISIS INTERVENTION AND NEGOTIATION**   **237**
Introduction   237
Types of Subjects and Situations   239
Characteristics of Negotiators   240
Dynamics of Crisis Situations   241
Dynamics of Hostage Situations   242
    Dynamic Factors   243
    Stockholm Syndrome   245
Crisis Intervention   246
Negotiation Approaches   249
    Criteria Measurements   252
Conclusion   253
Allen's World   254

Chapter 12   **POLYGRAPH USE**   **257**

Randy Walker, Retired FBI Agent, Walker Investigative Services
Introduction   257
What Is a Polygraph?   259
Purposes of Polygraph Examinations   260
    Pre-Employment Screening   260
    Personnel Security Screening   261
    Criminal Investigations   261
    Counterintelligence and Counterterrorism
        Investigations   263
    Misconduct and Internal
        Affairs Investigations   264
Polygraph Reliability   264
"Beating" the Polygraph   265
Computer Voice Stress Analyzer   267
Conclusion   268
Allen's World   269

Chapter 13   **STANCE-SHIFT ANALYSIS**   **273**

*Boyd Davis, PhD, Professor of Linguistics, University of North
Carolina-Charlotte, and Peyton Mason, PhD, CEO Linguistic Insights, Inc.*
Introduction   273
Statement Analysis   274
    Statement-Validity Analysis   274
    Criteria-Based Content Analysis   274
    Automated Techniques   274
Stance-Shift Analysis   276
Conclusion   281

PART IV   **INTEGRATION**   **283**

Chapter 14   **PUTTING IT ALL TOGETHER**   **285**

Introduction   285
Structuring the Practice Interview   285
    Identifying the Professional and Scheduling
        the Interview   286
    Developing an Objective and Corresponding
        Questions   287
    Conducting the Interview   287
    Transcribing the Interview   288
Analyzing the Interview   289
    Interview Objectives and Questions   289
    Establishing Rapport   290

Interviewer's Nonverbal Communication   290
Active Listening   290
Interviewee's Nonverbal Communication   290
Probing   290
Summary   290
Analyzing a Fellow Student's Interview   290
Conclusion   291
Allen's World   291

REFERENCES   **295**

INDEX   **303**

Preface

Overview of the Book

Interviewing in Criminal Justice: Victims, Witnesses, Clients, and Suspects is divided into four parts: preparation, conducting the interview, special areas of interviewing, and integration. Students will be exposed to the differences in investigative and helping interviews throughout the book as the authors balance the academic with the practical. Theoretical information is discussed and critiqued, but practical skills and techniques also are described. To help the students absorb, understand, and apply the information, the authors have included true-to-life examples to illustrate the information as well as exercises to practice the skills discussed.

Preparation   **Chapter 1** focuses on the foundation and process of communication. It is critical for effective interviewers to understand the process of how people receive—as well as send—information. The description of feedback and its importance to effective communications provides readers a foundation on which to build skills in delivering and receiving feedback (discussed in Chapters 4 and 5). An awareness of the factors that inhibit or enhance good communication allows interviewers to influence and enhance the outcome of interviews.

Chapter 2 focuses on the ethical issues surrounding interviewing, especially when the interviewees possess vulnerable characteristics. An effective interviewer is a powerful person, and he or she must remember to be ethical. The student should appreciate the power and responsibility of being an effective interviewer, as he or she will see in the variety of examples in Chapter 2.

Preparing for interviews requires substantial effort and time from the interviewer. A thoroughly prepared interview exponentially increases the probability of success. The better the preparation, the more accurate and thorough the information elicited will be. Because of its importance, two chapters are devoted to preparation (Chapters 3 and 4). **Chapter 3** focuses on the development of interview objectives and then the construction of relevant questions that meet those objectives. Most importantly, because the interviewer develops objectives for the interview, he or she knows with confidence if the interview is successful. Developing questions takes skill. The questions must be relevant. They must motivate the interviewee to provide information, and they must be constructed to enhance the interviewee's memory.

Conducting the Interview

Chapter 4 focuses on the mechanics surrounding the preparation of the physical setting and the crucial first few minutes of the interview. As part of the preparation phase, the interviewer should collect all relevant information about the interviewee. In addition, the interviewer should prepare the physical and verbal interviewing setting, which includes psychologically preparing him- or herself. Chapter 4 also extensively describes a number of fundamental communication skills needed to establish a connection with the interviewee, such as listening actively and seeking and receiving verbal and physical forms of feedback.

Chapter 5 focuses on the delivery of questions, evaluation of the interviewee's responses, follow-up probing, and feedback. The motivation of the interviewee begins with the wording of a question and continues with feedback from the interviewer during the interviewee's response. The interviewer's evaluation springs from the interviewee's verbal and nonverbal responses. As the interviewers observe interviewees' nonverbal cues and listen to their verbal responses, the interviewers assess the validity, relevancy, and completeness of the responses. Effective probing is an important tool that the readers will learn to facilitate these responses.

Chapter 6 focuses on documentation. Criminal investigations require written statements from victims and witnesses. Ideally, the detectives will obtain a confession, but, at a minimum, they must produce a written statement that they create and get the suspect to sign, or that the suspect creates and signs. Certain protocol should be included for statements that might be used in court. This chapter describes techniques that will encourage reluctant witnesses and suspects to write and sign a statement.

Interviews conducted by probation officers and in helping-intervention sessions also must be documented. In these interviews, the criminal justice professional writes the document rather than the interviewee. Chapter 6 also includes an overview of cutting-edge research on investigative discourse.

Specific Interviewing Issues and Techniques

Although this text is focused on the basics of interviewing, there are several other important areas covered. Specific interviewing issues and techniques addressing victims and children as well as different cultures are discussed in separate chapters.

Chapter 7 focuses on research recently conducted regarding victims' and children's needs. Victims, especially children, have special needs, and understanding additional techniques is critical when conducting these interviews. An overview of the theories describing children's developmental stages provides a foundation for comprehending the unique issues surrounding the interviewing of children.

Chapter 8 focuses on the importance of recognizing cultural differences during interviews. The literature on international communication is vast, but this chapter focuses on and exposes readers to an overview of these differences. The limitations of this text do not allow extensive discussion of specific cultures; however, there are general issues that should be understood and universal strategies that remain effective regardless of the culture. The authors provide a variety of examples and exercises in this chapter that will increase readers' sensitivity and appreciation of cultural diversity.

Chapter 9 focuses on the steps and techniques utilized by private investigators working with criminal defense attorneys. Although many of the basic interviewing techniques and strategies are used in preparing defense cases for plea bargaining, settlement, dismissal, or trial, there is sufficient variation to require a separate chapter.

Chapter 10 focuses on the use of interrogations. The emphasis of this text is on interviewing so that readers obtain the basic knowledge and skills needed to conduct an effective interview. Nevertheless, the authors provide an overview of interrogations here because it is particularly important for readers to be aware of the legal circumstances surrounding interrogations and the research critiques surrounding some of the controversial methods used in interrogation.

Chapter 11 focuses on crisis negotiations and intervention. Although crisis negotiations and interventions require more skills than simple interviewing knowledge, many of the same techniques and skills used in interviews apply to crisis situations. As with all of the specialty chapters, readers with an interest in these advanced areas are encouraged to expand their learning further.

Chapter 12 focuses on measures of deception. Here, the authors provide expert discourse from Randy Walker, a retired FBI agent and a skilled polygraph examiner. Walker walks the readers through an exam and explains the technology and the appropriate time and place for using a polygraph, helping readers to become better consumers.

Chapter 13 focuses on stance-shift analysis, an innovative and scientific advancement of investigative discourse described in Chapter 6. Researchers Boyd Davis and Peyton Mason describe the systematic means they use to analyze and interpret statements.

Integration   **Chapter 14** helps the beginner interviewer plan a practice interview and provides a structure for analyzing the practice and future interviews. Becoming an effective interviewer involves conducting many interviews and discovering what works and what doesn't work. Interviews can be conducted hundreds of times and still result in mediocre outcomes. As much as possible, each interview must be prepared and upon completion, each interview should be analyzed, asking the key question, "What could I have done better?"

Conclusion

The authors of this text have more than 70 combined years of interviewing experience. Vivian Lord has law enforcement experience and has spent much of her professional life teaching communication and interviewing techniques to law enforcement officers and college students. Allen Cowan has an extensive background digging for information, first as an investigative reporter for *The Orlando Sentinel*, *The St. Petersburg Times*, and *The Charlotte Observer* and then as a private investigator since March 1989. Given the authors' rich and varied backgrounds, each chapter is a blend of the theoretical with the practical.

Interviewing skills cannot be learned simply from reading a book. The authors can provide the techniques and strategies, but the skill and the effectiveness of any interview must be learned and practiced in small components. Once all of the elements are mastered, they must be synthesized and practiced on "safe" interviewees before incorporated into professional contexts. Students are encouraged to receive feedback on as many of the exercises included in this book as possible.

Acknowledgments

Interviewing in Criminal Justice: Victims, Witnesses, Clients, and Suspects is the result of the professional collaboration of several people at Jones and Bartlett. We wish to express our gratitude to Sean Connelly, our acquisitions editor, who realized the importance of this text, and our associate editor, Megan Turner, whose expertise and diligence were essential to bringing this text to publication. Credit also goes to Randy Walker (formerly of the Federal Bureau of Investigation, now of Walker Investigative Services), Dr. Boyd Davis (professor, UNC Charlotte), and Dr. Peyton Mason (Linguistics Insight, Inc.) for contributing their expertise to two special chapters.

A final debt of gratitude is owed to the professionals who gave the authors interviews, comments, and experiences that have enriched their lives to incorporate into this text.

We also wish to thank the reviewers of this book: Jeff Bumgarner, Minnesota State University; Christian Dobratz, Minnesota State University–Mankato; and James Pleszewski, Rowan-Cabarrus Community College.

Introduction

Professionals in the criminal justice field spend a great deal of time interviewing. The purposes of interviewing include obtaining information, giving information, solving problems, and helping to formulate options and make plans.

Because people begin talking as toddlers, they believe they are good communicators. Frequency of practice does not translate into effective communication. Unless their parents learned to be successful communicators, individuals rarely receive training in effective communication skills. As critical as communication is, it is rarely taught.

It takes training to be a good communicator, along with keen observational skills, empathic sensitivity, and focused attention. Reading the information and examples in this text, performing the exercises, and receiving feedback from trained instructors will provide readers with a solid foundation so they can become effective communicators.

Interviewing Defined

An interview is a conversation with a purpose between two people. Although there may be more than two people involved, an interview is dyadic: two people asking and answering questions. One person takes responsibility for the development of the conversation and guides its direction. The motivation of the two parties involved in the interview can be considered on a continuum, ranging from those who have little need or desire to be in the interview to those who have a great need (Shearer, 2005).

In some investigative interviews, there is mutual need between victims and investigators. The victim and the investigators share the common goal of solving the case. The victim has been injured either personally or through loss of property and is heavily invested in the identification and prosecution of the perpetrator. Also, financial compensation to help the victim recover often is a common goal.

On the other end of the continuum are witnesses with no personal relationship to the victim and suspects. They have little need or desire to be interviewed, while the interviewer has a great need for information. After all, the interviewer is seeking information that the witnesses and suspects are presumed to possess. The interviewer needs their information to close the case successfully after identifying, prosecuting, and convicting the defendants.

In counseling, social service, or probation interviews, the interviewer has the lower need for the interview and the interviewee has the higher. From the information gathered during these types of consensual interviews, the interviewer often makes decisions that impact the interviewee's life. For example, if one of the conditions of probation is employment, but the probationer has not become or remained employed, the interviewer, or probation officer, will conduct an interview with the probationer to discover the factors that have hindered the probationer's ability to get or keep a job. If the probation officer believes that the probationer has not made an effort to find a job or has demonstrated poor work habits that led to losing the job, the probation officer has the authority to take action. The officer can take the findings to the court, stating that the probationer, having failed to meet the terms of probation, should have probation revoked and serve out the terms of his or her active sentence.

What is one difference between investigative interviews and helping interviews? Motivation. In investigative interviews, with the exception of most victims, the interviewer is highly motivated in obtaining information, while the interviewees, witnesses, or suspects have little to gain from the interview. In helping interviews, the interviewee usually is highly motivated to get help (unless court referred).

Interrogations, on the other hand, are dominated by the interviewers. Their purpose is to coerce and cajole. The purpose for the interrogators and their subjects are definitely not mutually beneficial, but rather direct and intense. There is a great deal of controversy surrounding specific interrogative techniques. Some of the ethical issues surrounding interrogation are discussed in Chapter 3. As a basic text on interviewing, it is not appropriate for this text to discuss interrogation extensively. Chapter 11 does provide an overview of psychological theories of deception and interrogation, legal parameters surrounding interrogations, interrogation approaches, and issues surrounding the approaches.

Criminal Justice Positions and Interviewing

The most obvious use for interviewing is by law enforcement. Line officers—and then more specifically, detectives—must obtain pertinent information from the victims of a crime or those individuals who may have witnessed the crime. The consequences of police interviews are serious. Without complete and valid information that can be documented, crimes may not be solved, or an innocent person may be apprehended. Incarceration is a serious consequence.

Many of the law enforcement interviews will be conducted with reluctant interviewees, who perceive the professionals as adversaries. It takes experienced and trained interviewers to acquire valid and reliable information through thorough investigations, reinforcing the need for preparation before conducting the interviews.

Interviewing is essential for decisions made in court. Defense attorneys with their accompanying private investigators will parallel the police investigations to adequately defend the alleged perpetrator. Witnesses who are friendly toward the prosecution will be reluctant to talk to defense investigators. Although defense investigators are gathering information for their client's defense, the defendant may not necessarily provide valid and complete information. In the "Allen's World" features found throughout the text, readers will learn strategies used by an experienced private investigator to obtain complete, relevant, and valid information from all parties, including witnesses for the prosecution.

Judges also need information for sentencing. Probation officers will be involved in compiling risk-assessment information for sentencing. The thoroughness of these interviews is critical to decisions that protect the public from violent offenders but also to consider the defendant's potential for rehabilitation. Obtaining valid information from family members, employers, teachers, and friends gives the judge a better picture of the defendant's probability of success if given probation and/or referred to alternative-to-incarceration programs. Deferred prosecution may be an option, but a judge must understand the risk of letting a convicted person roam the community.

The growing number of alternative programs uses helping interviews in efforts to facilitate their clients' success in becoming law-abiding citizens. The staff at these programs helps offenders manage anger, cope with stress, and communicate effectively with a variety of people in their lives. During interview sessions, interviewers help the interviewees explore and express feelings. Attentive listening by interviewers is used to understand and help the interviewees appropriately verbalize their feelings. In other words, the message intended must be the message received.

During the initial interview with clients, information is provided. The interviewer explains conditions of probation, the program, or jail rules. One of the components of interviewing is responding to questions and ensuring understanding through feedback. Because negative feelings may be expressed toward the interviewer, interviewers must learn to deflect any hostile words and to remain objective. Interviewers must understand the extreme emotions their clients may be experiencing when they face jail and/or probation, thereby making the clients' ability to listen difficult. Only interviewers who learn how to reflect feelings and content will be able to decrease emotion and increase rational thinking needed by clients to listen effectively.

It then becomes mandatory that criminal justice professionals, working with clients on probation or in alternative programs, include problem solving and planning as a component of their helping interviews to facilitate the goals of their clients regaining their place in society. The interviewer must develop flexibility to move from investigative interviewing skills to helping interviewing skills, always mindful that the client may be reluctant.

Professionals working with victims gather information and conduct helping interviews to assist the victim's return to a normal life. Skills in crisis intervention become particularly critical for helping victims of personal crimes. Coping with a crisis such as robbery or rape often is unmanageable and will leave victims feeling completely helpless and fearful to even deal with daily life.

Juvenile court counselors and treatment personnel working with troubled young people have the additional challenge of developmental issues. Physical, mental, and emotional elements differentiate children at different ages and children from adults. Specific communication challenges must be overcome to obtain complete and valid information from young people.

In all of the different criminal justice fields, professionals must be prepared to interview a rapidly changing population. The United States has become, and will continue to become, more diverse. There are general interviewing techniques, but interviewers must be prepared to distinguish differences in cultures; it is dangerous to generalize too broadly among Asian or Latin American cultures. The more interviewers can learn about specific cultures they will be regularly interviewing, the more likely that they will be able to elicit and receive valid and complete information.

Although some of the information for criminal justice responsibilities comes from written information such as court files, a large proportion of it originates directly from people, fragile human beings who will be reporting what they think they saw, what they remember, what they believe, or what they value. All their communication is filtered through their own particular communication style, cultural background, mores, and a variety of other elements.

PART

1

Before Interviewing

Foundations of Interviewing

Upon completion of this chapter, students should be able to

1. Apply basic communication theory to daily communications and interviewing
2. Become aware of inhibitors of effective communications and the use of feedback
3. Utilize the principles of interviewing
4. Connect basic human needs to effective interviewing

Basis of Communications

Communication is a symbolic process in which people create shared meanings (Lustig & Koester, 1999). Humans communicate through the written word, verbally and nonverbally. Words are formed as symbols to represent concrete images and abstract thoughts and beliefs.

A mother makes eye contact with her baby. Pointing at herself, the mother says "mama." At some point, the baby mimics her mother and says "mama." The mother communicates her excitement and pleasure at the baby's word; she displays positive reinforcement, so the baby repeats "mama."

The baby's young mind does not comprehend that the word *mama* symbolizes the person holding her. It will be a year or more before the baby's mind matures to the degree it can link the symbol *mama* to mean the baby's caretaker. Linking concrete images that we can observe to abstract symbols or words is an advanced level of brain development.

Even an animal as concrete as a dog is described in a variety of ways based on the varied experiences of different people. The simple word *dog* conjures up memories of personal pets, television heroes like Rin Tin Tin or Lassie, or movies such as *Beethoven* and *Marley and Me*.

The word may not always be pleasant. A person may have been bitten by a dog, knocked down by a dog, or seen a movie in which dogs were the aggressors. Almost all words, no matter how simple they seem, such as *dog, boy, juice, road,* and *land* invoke different images to different people.

When individuals attempt to discuss abstract thoughts such as criminal offenses, penalties, restitution, and justice, there is nothing directly to see,

EXERCISE 1.1

What Are Words?

Activity: The instructor asks different questions that reflect confusing areas.

Purpose: To facilitate discussion of words and what they represent.

1. The instructor asks different students to think about a dog and then to describe it. The students are likely to have a variety of descriptions with different sizes, colors, types of tails, ears, and length of hair. The instructor asks the students what they are basing their descriptions on, e.g., their pets or dogs in movies or television shows. It is helpful to stress the idea that none of their dogs are the right dog, but rather what their past experiences recall.

2. The instructor can come up with a number of confusing phrases to dictate to the students. Ask the students to write their responses on the board and then discuss what they meant to the students.

   Some examples: ship sails today (ship sales today)
   sons raise meat (sun's rays meet)
   let's blow this joint (leave a location or explode a site, smoke marijuana)
   ramp (exit off a highway, board connecting to areas, onion-like plant that is eaten in the Blue Ridge Mountains)

hear, touch, or taste. Humans must take a variety of experiences from parents, school, friends, and other institutions and create what they relate to abstract words. These creations, or concepts, are vastly different among humans, and only if they make a conscious effort to relate during conversations, watching movies, reading, or interviewing, will they be able to understand each other.

As we will discuss in Chapter 4, nonverbal communication can drastically change the interpretation of the words transmitted. The speaker may be presenting words that are neutral or even positive, while his or her nonverbal cues are transmitting negative thoughts. An individual with clenched fists, a red face, and bulging eyes who is snarling loudly, "Nothing you do will bother me!" is not likely to be believed. All of his or her nonverbal cues are communicating anger and outrage even while the words deny it.

What is particularly risky in the serious area of criminal justice is when people assume they understand, and that they are being understood. There are multiple reasons that mutual comprehension may not be reached, but without feedback, those involved in the discussion or the interview believe they do understand or they are understood.

How many people have been in arguments with their loved ones and said, "That is not what I meant. You don't understand at all!" If it is common for us to misunderstand our significant others, how much easier is it to misunderstand strangers? Criminal justice professionals must remain sensitive to the need to take the time and effort to ensure they understand what is being told to them. Equally important, professionals must use feedback to ensure they are understood by the citizen, client, victim, or inmate.

The following is historically factual. It illustrates how simple, seemingly easy-to-understand words can have fatal consequences.

In 1951, two young British boys, Derek Bentley and Chris Craig, were caught by the British police breaking into a store.

Bentley managed to go into hiding. Craig, armed with a pistol, was left out in the open, face to face with Constable Sidney Miles.

Miles ordered Craig to drop the gun.

It has never been proven, but it was alleged that Derek yelled out, "Let him have it, Chris." Words uttered during a tense, emotional confrontation with no time for either party to get clarification.

Did Chris think that "Let him have it" meant his friend, Derek, wanted him to shoot the officer? Or, did Derek want Chris to turn over the gun and surrender? These two questions have been debated ever since.

Whatever the real answer, Chris fired. Constable Miles fell over dead.

Derek quickly went on trial.

The Crown argued he intended for Chris to fire at the officer.

The defense argued that Derek intended for Chris to surrender.

Despite the fact Derek was epileptic and retarded, he was convicted and executed.

Chris, 16 at the time, was tried as a juvenile offender. He went to prison and was paroled after serving 10 years.

On July 30, 1998, Derek was posthumously pardoned. Two deaths. Tragic results from miscommunications. Did any good come out of the trial and its aftermath? The British abolished the death sentence. ('Let Him Have It!'—The Case of Bentley and Craig, 2006)

Communication as a Process

Communication is a process, changing and evolving with distinct, but interrelated, steps. Each person is part of the process and becomes a source who formulates and expresses or encodes ideas in ways that are consistent with his or her orientation or belief system.

Through encoding, the speaker selects and arranges words according to rules of grammar applicable to the language used by the speaker. The encoding leads to the production of a message, which is a set of verbal and nonverbal symbols.

The person who is channeling the message, through sight or sound symbols, interprets or decodes in terms of his or her orientation, and then encodes his or her response. In this way, the orientation of the individual changes the meaning of the communication that is sent to him or her, but it also is possible for a message to influence the orientation of the person receiving it. Through this ongoing process, communication involves shared meaning.

McMains and Mullins (2006) add an additional element to the communication process that they label *noise*—defined as anything that interferes with the message.

Interference can be background sounds or poor transmission that includes static on the telephone lines. Noise also can be perceptual. Interpretation by the encoder and the decoder are based on the parties' orientation, or their beliefs, age, education, and experiences. Unintended meaning can be assigned by either the encoder or decoder due to differences in their orientation. Psychologists call this *filtering*. Individuals use words to convey messages based on their experiences and what those words have come to mean to them. Receivers interpret the words they hear based on their experiences and what those words have come to mean to them. At times, this leads to confusion and misunderstandings.

EXERCISE 1.2

Communication Model

Activity: Students are told to think of one person with whom they have the most trouble communicating. They are to list at least 10 traits of that individual.

Purpose: To comprehend the influence of different beliefs, experiences, and values on individuals' ability to effectively communicate.

The instructor asks the students to think of a person with whom they currently, or in the recent past, had the most problems interacting with. The instructor should not place any criteria on who they select. The students then list a minimum of 10 traits of that person on a sheet of paper. The students can think about personal characteristics such as gender, age, education, any military experience, state or country of origin, as well as more abstract traits such as belief system, professional background, or values. Students are then requested to volunteer a description of the individual (without naming the individual) including some of their listed traits. The instructor then asks the students:

1. How do the traits of your described individual differ from yours?
2. How do you think those traits influence how that individual communicates with you?
3. How do you think those traits influence how you communicate with the individual?

For example, a 60-year-old businessman with military experience articulates his thoughts based on his interactions with people from his business and military background. If he is addressing or interviewing an 18-year-old female student, she screens his words through her youthful gender values and norms. Her world does not include either his military experience or his business background. Likewise, the military man-turned-businessman will have difficulty relating to the young female student. They come from two different worlds.

As noted in **Figure 1.1**, much like light is refracted when mirrors are slanted, communication between two people is almost always distorted. The more different the orientations of the two people, the more likely there will be confusion and misunderstanding unless both parties are aware of their differences and work diligently to clarify the message through feedback.

Figure 1.1 The Communication Process

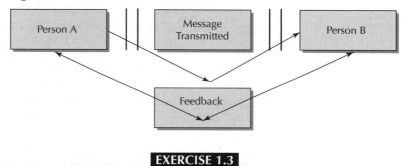

EXERCISE 1.3

Domino Feedback

In triads, the students will describe their dominoes configurations using different amounts of feedback.

Purpose: To help students comprehend the importance of feedback.

The instructor will divide the class into groups of three. Each group will be given two sets of the same six dominoes. Two students will sit back to back with the third student observing. The students will carry out the exercise three times with different instructions. The instructor should allow the students to rotate. Each time the A player will make the configuration, and the B player will attempt to make the exact same configuration. The C player will observe. At the end of each round, the C players will give their observations to the class.

Round 1: The A player will build a configuration with the six dominoes. It is not necessary to follow dominos rules; the configuration can be laid down flat or built up. The dominoes should touch each other in some fashion. Once the A player is finished, he or she will describe the configuration to the B player. The B player is not allowed to say anything. Once the A player has completed the description, the two players will turn around and see how closely their two configurations match.

The C player should then tell the two players what he or she observed. All the C players will then share with the class.

Some of the areas to be considered: (1) How complex was the configuration? (2) Did the A player use descriptions to help orient the B player, for example, "take the domino with the two dots and four dots and place it flat so that the two dots are pointed toward the back wall of the classroom"? (3) Did the A player remember that the description would be the mirror image, e.g., player A's right hand is player B's left hand? (4) Was A sure that B would have the exact replica? If B could have asked questions, would they have had a more exact match?

Round 2: A builds a configurations and then states, "ready." B then begins to ask questions to make the same configuration. A is able to provide complete answers, but does not ever actually describe the configuration nor initiate new information. The only answers provided will be in response to B's questions. Once player B believes he or she has the same configuration, players A and B will turn around and view the two configurations. Again, the C player shares his or her observations including addressing the questions above.

Round 3: Once A builds a configuration, A will begin to describe the configuration. B is allowed to ask questions at any point. A also should take time to confirm verbally what B is building. Both players should attempt to provide specific information and ask clarifying questions. Only after A and B believe they have the same configuration, should they turn around and look. When C relays observations, he or she also should add what descriptions or questions seemed to be most helpful or ineffective, and what questions should have been asked that were not asked.

Rumor

Purpose: To demonstrate the ease in which messages become confused without feedback.

Most students will remember the children's game of "rumor." One person whispers a brief message in the ear of the person beside him or her. This person in turn whispers what was said to the next person. Once the message has been relayed throughout the class, the last person repeats the message out loud. The message often is drastically distorted.

Only through feedback within two-way communications can the accuracy of meaning be assured. Feedback is the process of correction through incorporation of information about the effect of one's performance (Yeschke, 1997, p. 46). Ideally, both parties listen carefully, then rephrase, or mirror, what they believe they heard, and ask questions if there is confusion. The other party confirms or rewords. It is critical that both parties understand that neither is right. Rather, they must understand that both of them are working to deliver and receive the words that can be understood accurately.

Communication is intentional and unintentional behavior. The speaker intentionally attempts to convey the meaning of a message to another person. The speaker is consciously selecting words and meanings to persuade the other person. At times, communication also is unintentional. Some argue that communication takes place whenever meaning is attached to another person's actions (Samovar & Porter, 1991).

While communication must be dynamic, it is also irreversible and irrevocable. As all people have regrettably experienced, once said, words cannot be taken back. Humpty Dumpty could not be put back together, and toothpaste cannot be put back into the tube. While they may not "break our bones," it is not true that "words can never hurt us."

Inhibitors of Effective Communications

There are differences between hearing and listening.

When a human or any animal collects sound waves at different frequencies, which are physiologically transmitted via electrochemical signals to the brain, the animal is hearing. Listening, on the other hand, is the psychological process of ascribing meaning to those signals.

There are, as we all know, obstacles to effective listening. Differences in word meanings and differences among individuals have been previously discussed in this chapter and can create major problems. That is one reason why the following two types of speech should be avoided.

The first is *polarization*, phrasing everything as good or bad. Such a rigid, inflexible way of seeing the world will put up a wall between two or more people who cannot see middle ground or the gray between black and white.

The second type of speech to be avoided is the use of the word *all*; statements that overgeneralize are communication stoppers. Once a citizen or young person hears, "all teenagers are bums just waiting to get in trouble," effective communication shuts down (McMains & Mullins, 2006).

There are other obstacles to effective communication that should be discussed, and if possible, eliminated. The most obvious obstacles are physical and environmental. Noises from the surrounding area should be controlled by finding a quiet location to conduct an interview without noise caused by traffic, machines, cellular phones or the ringing of other telephones, television, radios, or conversations. Uncomfortable seating or room temperature also should be considered. Lighting will be discussed more in Chapter 4; however, hot, bright lights are for movies, not real life.

Other physical obstacles may be more personal. The interviewee may have a hearing impairment or language difficulties. Even if English is the interviewee's first language, limited vocabulary comprehension or capabilities hinder the interaction between the interviewee and interviewer.

Professionals need to resist the use of jargon, idioms, or words from another language that have crept into the lexicon. Jargon is only understood by others in the profession and easily adds to misunderstanding. **Figure 1.2** presents a well-known joke that reflects misunderstanding culminating from the use of jargon. A police officer who yells, "Get 'em up," may believe the message is clear. The officer who issues that command knows what it means. The offender is to put his or her hands up; however, individuals who have had

Figure 1.2 Make Sure You Are Understood

A colonel issued the following directive to his executive officer:
  Tomorrow evening at approximately 2000 hours, Halley's Comet will be visible in this area, an event which occurs only once every 75 years. Have the men fallen out in the battalion area in fatigues, and I will explain this rare phenomenon to them. In the case of rain, we will not be able to see anything so assemble the men in the theater, and I will show them films of it.

Executive officer to company commander:
  By order of the colonel, tomorrow at 2000 hours, Halley's Comet will appear above the battalion area. If it rains, fall the men out in fatigues, then march to the theater where the rare phenomenon will take place—something that occurs only once every 75 years.

Company commander to lieutenant:
  By order of the colonel, be in fatigues at 2000 hours tomorrow evening. The phenomenal Halley's Comet will appear in the theater. In case of rain in the battalion area, the colonel will give another order—something that occurs once every 75 years.

Lieutenant to sergeant:
  Tomorrow at 2000 hours, the colonel will appear in the theater with Halley's Comet—something that happens every 75 years. If it rains, the colonel will order the Comet into the battalion area.

Sergeant to squad:
  When it rains tomorrow at 2000 hours, the phenomenal 75-year-old General Halley, accompanied by the colonel, will drive his Comet through the battalion area theater in fatigues.

little exposure to the legal system may interpret the phrase as moving other people, moving articles to another area, or picking up a child or another person. The officer who feels in danger may then become scared and possibly lethal.

Of course, if the individual is under the influence of alcohol or other drugs, communication becomes impractical. Safety requirements suggest that, rather than talking to inebriated individuals, it is better to confine them in a safe environment until they become sober.

In addition to language differences, cultural differences may cause confusion or build a barrier between the parties. Differences in nonverbal cues can be emotional sources for misunderstanding. As we will discuss in Chapter 8, in-depth study into the more common cultures and ethnic groups in a community can prevent many of the communication problems that arise when interviewing or interacting with people from a different culture.

For example, in the United States, a dinner guest might bring red roses as a friendly gesture. In Germany, bringing red roses to a hostess will be interpreted by the recipient that the guest has feelings toward her. Also, it is common in the United States for guests to admire objects in a home they are visiting. In Germany, if the guest admires an object, the hostess has been conditioned that it is good manners to give the object under admiration to the admirer.

Less obvious obstacles to understanding are psychological barriers, or any mental or emotional conditions. The average person thrust into situations with individuals in positions of authority may experience an increase in emotions, such as apprehension or anxiety, that could inhibit that individual's ability to listen. The possibility that one of the parties to a conversation or interview is undergoing emotional distress also must be addressed before listening can effectively occur. Like a teeter-totter, when emotions are high, rational thinking (and listening) is low. Allowing the interviewee to ventilate, or express some of his or her emotions, helps develop more of an emotional balance so that rational thinking and consequently, listening, increase.

As will be discussed in Chapter 4, vigilant observation of the interviewee helps the interviewer discern if the interviewee is distracted or is having problems communicating. Feedback also is useful when attempting to remove barriers.

Human Needs as Motivators

Although researchers categorize human needs in a variety of ways, Abraham Maslow developed a hierarchy-of-needs model that Charles Yeschke (1997) concludes should be considered during interviews.

Maslow's hierarchy of needs lists three primary areas of human needs. First, physiological needs such as air, food, and water must be met before humans strive to meet the second level of needs—safety and security. Once safety and security are met, they then strive for the third level of love and belonging. Many people throughout the world struggle to meet these three levels and never attempt to fulfill the next two levels, which are self-esteem and self-actualization.

Physiological needs have limited importance during interviews; however, interviewees should be comfortable. They should be rested, offered something to drink, and be allowed to visit the restroom.

Safety and security are fundamental to successful criminal justice interviews. Because of the need to feel safe, interviewees seek to control their environments. Although on the surface, interviewers have power, interviewees may not necessarily submit and relinquish control unless they feel safe.

Interviewers must be aware that people want to belong and interact with others. They want to be cooperative. So, when they believe they are choosing to share information and not just submitting to the interviewer's authority, they are more likely to provide more complete and valid information.

Also, when people have done something wrong as defined by their social group, they fear being ostracized by that group. They will be reluctant to admit to an act for which they may lose membership in their primary group (their family), or an extended group. The number of sexual acts perpetrated by Catholic priests and teachers on young boys over the decades, but only recently coming to light, has horrified the public. The reluctance of the boys to come forward demonstrates the shame they felt as victims and their fear of being labeled as different. The denial by the Catholic Church of the sexual acts perpetrated by its priests amplifies the fear of loss of reputation of the individual priests and the lack of protection that the Church provided. All of these parties have denied out of fear of rejection by their social group.

Self-esteem is often translated to "saving face." Most interviewees, including witnesses and victims, are motivated to do what is necessary to defend their self-image. They are sensitive to responses from persons in the position of authority. They desire respect, understanding, and acceptance. Any response or gesture by the interviewer that might damage their sense of self will be confronted defensively.

Interviewees never have to lie or omit information. They have a choice, although it may not always be an easy one, especially if they are worried about the cost of their freedom, as well as their self-esteem. They do not want to be considered ignorant, immature, or uneducated. They react to how they are treated. If they feel accepted, or at a minimum, understood, then they are less likely to lie.

Interviewees need emotional room to maneuver. They may remember a detail they previously forgot, or they may decide to volunteer information when they feel more comfortable. The interviewer should leave the door wide open, not just ajar, so interviewees can add information without fear that they will be considered liars.

As interviewers are preparing their interviews, they need to have sufficient information about the interviewees to consider what basic needs the interviewees are seeking to have filled, or that they are defending. If the interviewee is a middle-age executive of a bank, belonging and then self-esteem are more than likely important needs for him or her to defend. The interviewer will need to carefully preserve the banker's self-esteem if an interview is to be completed without attorney interference. For others, it may be safety. A gang member who has agreed to turn on other gang members could be in physical danger—not only in fear of no longer belonging to a social group. The interviewer needs to be fully aware and prepared with this information.

Reducing Anxiety

There are a number of behaviors that individuals may exhibit if they are experiencing anxiety. Acute anxiety is more likely to be caused if the individual feels a need to deceive the interviewer. The individual seeks means to decrease anxiety. The more directly the individual must lie, the more anxiety he or she feels (Jayne, 1986). Therefore, the interviewee will use strategies to minimize direct lies in order to feel less anxious.

Omission is a common strategy used to minimize direct lies. Omissions, such as shaking or nodding the head, responding with silence, or responding with incomplete facts, are useful tools to minimize anxiety. Most young people can remember a time when they came home after curfew. When asked where they had been, they would detail their brief visit to the library but omit the extended visit to a friend's house. An individual who is in potential trouble with the criminal justice system because of an offense that occurred on a specific day, may describe everything he or she did that day except for the offense in hopes of diverting the interviewer's attention away from the offense.

Evasions, such as feigning disbelief or answering a question with a question, often cause slightly more anxiety. The interviewee is providing a distraction, but no real answer. For example, a cashier is asked if she took money from her cash register. Rather than answering directly, she displays shock and asks, "You think I stole money?" or "Why would I steal from the store?"

Another form of evasion is a qualified answer. A probation officer asks a client if he or she used drugs since the last appointment. A qualified answer would be, "I haven't touched crack since I started probation." This evasive answer might be interpreted that he or she has used no drugs, has used a different drug than crack, or smoked it without touching it.

Inbau, Reid, Buckely, and Jayne (2004) describe other evasive types of responses. These responses include poor memory such as, "as far as I know" or "to the best of my knowledge," or qualifiers such as, "not really" and "mostly." The phrase made famous by Jerry Seinfeld, "yada, yada, yada," or the newest in vogue evasive response, "whatever," are also examples.

The interviewee may attempt to reduce anxiety mentally through defense mechanisms in which reality is redefined to reduce anxiety, guilt, or loss of self-esteem. Rationalizations and projections are the two major defense mechanisms.

Rationalization invents justifications for actions. These explanations are attempts to avoid responsibility or intent for behavior. Interviewees will protect themselves with rationalizations if they have images of themselves that are not supported by fact (Yeschke, 1997). Sykes and Matza extensively describe rationalization with their "techniques of neutralization" (as cited in Vito, Maahs & Holmes, 2007). They categorize rationalization, or criminal-thinking errors, into denial of responsibility, denial of injury to the victim, denial of an actual victim, condemnation of those who dare condemn, and appeals to a higher loyalty such as family or gang. These different categories all provide interviewees means to justify their behavior. If the interviewee cheated on his or her income taxes, the categories of denial of injury to the victim, denial of an actual victim, and condemnation of those who dare condemn are all plausible. After all, who is he or she cheating, and if it is the U.S. government, who has been hurt? The U.S. government is not helping the interviewee get a job or make his or her life any easier.

In another example, an accountant may have *borrowed* money rather than embezzled. The accused accountant intended to return the money, but rationalizes that the *loan* was discovered before he had a chance. A pedophile rationalizes that he is simply giving attention to a neglected child—a deprived child who has never received any gestures of love from parents. To the pedophile's way of thinking, he is giving the love that child needs.

Projections, on the other hand, shift the blame for actions or behaviors onto another person. A domestic-violence batterer may project blame for his or her assault onto the victim. If the victim would quit nagging or fix dinner on time, the batterer would not be forced to hit or slap her; she had it coming. An offender of white-collar crime may blame his or her employer for refusing to pay overtime. As an employee, he or she is just getting what is due from all of the hard work.

Rationalization and projection are common. As noted by the self-serving attribution theory, characteristically people give credit to themselves for their successes and blame others or the situation for their failures. Such actions allow them to feel better about themselves and to attempt to make themselves look better (Wetzel, 1982).

Principles of Interviewing

The complexities of human nature and communication can be simplified with basic principles of interviewing that can also be used as life lessons for belonging to a society consisting of other humans (Samantrai, 1996; Yeschke, 1997).

Individualization   Generalizations about people of certain ages, ethnicity, or socioeconomic groups have no place in interviewing. Every person has unique qualities and should be interviewed without bias, prejudice, or judgment by the interviewer. Empathy, the willingness to understand the individual and his or her feelings, is an important component of active listening and is a skill fully developed in Chapter 5.

Curiosity   Every individual and situation must be perceived as interesting. Interviewers are fact gatherers, not judges. Interviewers should meet interviewees with an open mind and be prepared to fully gather information without preconceived ideas.

Self-awareness of their own inhibitors of effective communication will facilitate the interviewers in analyzing each answer for completeness and relevancy to the objectives of the interview. Interviewers must be aware of words that set off emotional triggers causing expressions of surprise or disgust to be expressed on their faces. Developing means to detach emotionally from these triggers comes with self-awareness and interviewing experience.

Flexibility   Effective interviewers should prepare to interview by extensively collecting background information about the interviewee and the related incident, identifying the purpose of the interview, and developing questions before conducting the interview.

This preparation permits the interviewer to be attentive during the interview and allow for flexibility if the purpose of the interview is not fulfilled, or if an answer opens a tangential line of questioning that might be beneficial to follow. Based on the individuality of people mentioned earlier, it is important for the interviewer to respond individually. Learning to improvise probes during the interview, as discussed and practiced in Chapter 5, is a critical skill.

Self-Fulfilling Prophecy

Interviewers need to remember that expectations we have of others and how we behave toward others influences their responses. If the interviewer enters the interview believing the individual is a liar and of little use to society, these beliefs will be reflected in the interviewer's tone, expressions, and nonverbal communication. In turn, the interviewee will read the negative thoughts of the interviewer and respond negatively.

On the other hand, if the interviewer behaves positively, then the interviewee is more likely to respond positively. If the interviewer believes the interview will uncover important information and the interviewee will be cooperative, then more than likely the interview and interviewee will fulfill that positive prophecy. The interviewer's tone, expressions, and nonverbal cues will communicate the positive beliefs.

Self-Confidence

Interviewers also need to develop a positive set of expectations for themselves. With training and practice, they know they can become effective interviewers. Self-confidence is important and will exude through every pore. Self-confidence is not to be confused with arrogance or false pride; it is based on hard work and training. Self-confidence keeps interviewers centered and able to remain objective. Interviewers lacking self-confidence are more likely to become defensive and to bully interviewees.

The interviewers should visualize the interviewer they would like to be.

EXERCISE 1.4

Galatea Effect

Activity: The instructor takes the students through a visualization exercise.

Purpose: To help the students build self-confidence in their ability to conduct interviews.

Preferably at the end of a class session, the instructor tells the students that she or he will take them through a visualization exercise they can use at home. The students are instructed to get as relaxed as they can in their seats. They should close their eyes and place both feet squarely on the floor and their hands in their laps or on the desk. The following script can be read slowly and calmly or even taped.

"I want you to breath in slowly, holding your breath—1, 2, 3, 4 and then letting it out slowly—1, 2, 3, 4. Again, slowly breathing in—1, 2, 3, 4—and out—1,2,3,4. Look inward and see yourself rising from a restful sleep, relaxed and ready to start the day. You see yourself going about your normal morning routines. You dress in your most comfortable, but professional, attire. When you look in the mirror, you see a well-groomed, confident you looking back. You smile at yourself. You know that you are prepared and ready to conduct the interview you have

> planned today. You are prepared, know the background information, and have developed probing questions. You see yourself arriving at the location where you are going to conduct the interview. The room has been carefully set up. You are prepared. You are ready. As the individual you are interviewing enters the room, you approach him or her with confidence and casually look in his or her eyes. You know you will get the information you seek. The interview will have a positive outcome.
>
> Now, I want you to take a deep breath, breathing in—1, 2, 3, 4—and out—1, 2, 3, 4. Now, slowly open your eyes."

Expression of Feelings   The need for expressing feelings and sharing experiences with others is very powerful and works to the interviewer's advantage. This need is especially felt when a person is in some kind of trouble or difficulty, which intensifies the need for sharing the burden with another person. As noted earlier, if the interviewee's emotional level is high, then his or her rational thinking is low. Helping the interviewee express feelings moves the interview forward in a constructive direction.

Self-Determination   People have the right and the ability to make their own decisions and choices, although the ability may be at different levels of competency. A criminal justice professional working with a client is fostering a relationship that will stimulate the client's inner strengths. The goal is for the client to become an independent and productive citizen.

Confidentiality   The right to privacy is not absolute. Legally and ethically, the interviewee needs to know what information can remain confidential and what information will need to be shared and with whom. When possible, he or she needs to feel safe to talk freely about problems and to know the interviewer will not misuse the information.

Conclusion

Chapter 1 provides readers with the foundations of communication. Understanding the complexities of communication helps interviewers appreciate the importance of feedback and the significance of two-way communication and understanding. There are a variety of barriers to effective communication, and every attempt must be made to eliminate those barriers if there is to be a successful interview. Likewise, comprehending the responses to anxiety also facilitates the interviewer's interaction with interviewees who choose to use such techniques as rationalization, projection, or omission.

Effective interviewing skills are based on basic communication skills and understanding human needs. Interviewers who see themselves as fact seekers and believe in the worth and dignity of those they interview will discover they are able to obtain valid information and to remain human themselves.

ALLEN'S WORLD

Note: Throughout this book, names have been changed, but the situations are real.

I first met Mike in the Mecklenburg County jail.

The room is about 6 feet by 10 feet. The walls are concrete block, painted beige. Mike sits on one side of a bullet-proof glass. He is wearing an orange jumpsuit. I am on the other side of the glass. We talk through a small opening in the glass.

I tell Mike who I am. I point out that at the moment, and for the foreseeable future, I am the most important person in his life—not his attorney or the prosecutor—me. I am his fact finder.

Mike wants to talk about his family.

I do not.

Mike wants to chit chat.

I do not.

I tell Mike the importance of this initial interview.

Many of the techniques Dr. Lord describes are not relevant when I am interviewing a defendant. I do not particularly care if Mike likes or respects me. I am gruff. Some might say harsh.

"Mike," I say to him at the outset, "if you ever lie to me, you'll never see me again."

That's how I get his attention. And it works.

I tell Mike that I am working for, and employed by, his attorney. There are two reasons for this. One, I want to make sure I get paid. If Mike is convicted and goes off to prison, he has no incentive to pay me. The other is for legal purposes. As long as I am working for an attorney, what Mike tells me and what I uncover during my investigation is privileged information. Mike must know that even if he confesses, his conversations with me are confidential.

I rarely ask a client, someone like Mike, if he or she committed the crime he or she is charged with. I do not care. We do not have to prove Mike did not do something. The state must prove, beyond a reasonable doubt, that Mike committed the murder. There is a chasm of difference.

Extensive preparation is the key at this juncture. I have reviewed the file compiled by his attorney, which includes Mike's original interview with the attorney. I have read the indictment. I have almost memorized the discovery provided by the prosecution.

Among other charges, Mike is facing a murder charge.

When his story seems thin, I challenge him. I tell him it does not comport with what is in his file—or the indictment. Mike backtracks, a lot. But, in the end, I get what I need out of this initial interview to help his attorney prepare a defense. (Ultimately, Mike walked on the murder charge but did serve time in prison for unrelated drug charges. He was released in late 2008.)

Just as interviewees rationalize, so do I. It is a way of making myself feel better, and it allows me to effectively do my job. I rationalize a lot to help my clients. I tell myself that once someone injures my client, there should be no restriction on me in defense of my client.

For example, an individual's right to privacy is less important to me than my client's defense. I can get cell phone records. From them, most of the time I can find out what towers relayed the calls and where the person was when those calls were made. It is an easy way to confirm an alibi or to impeach a witness.

I can get financial information. What better way to find out if a suspected embezzler made an unexpected, large deposit shortly after the money was embezzled?

I follow people. I might want to find out if they are filing a fraudulent workmen's compensation claim, or cheating on a spouse.

I do whatever my conscience will allow in defense of my client. Am I rationalizing immoral behavior? As I tell students when I lecture, it is my job.

Ethics Surrounding
Interviewing

OBJECTIVES Upon completion of this chapter, students should be able to

1. Understand the ethical issues and their ethical responsibility surrounding interviewing
2. Be able to develop their individual ethical standards
3. Understand the potential power of their role as interviewers and the potential consequences of interviews

Introduction

A skeleton shadow of a woman sat across from the interviewer. She would not make eye contact, but showed no emotions. Her infant was dead, allegedly from Sudden Infant Death Syndrome, but there were indications that the child may have died from neglect and abuse. The interviewer knew that it would be tempting to treat this woman with scorn and ridicule. Any person who would put his or her drug habit before an infant's needs did not have any self-respect and deserved to be treated with scorn.

Ethical behavior is based on knowing the difference between what is legal and what is moral. The interviewer is solely responsible for the interview methods employed; his or her conscience acts as the benchmark.

Regardless of the alleged offense or how unsavory the personality of the interviewee, the interviewer must adhere to ethical standards. The interviewer controls his or her behavior; it is not controlled by the interviewee, the criminal justice system, or the immediate situation.

This chapter provides examples of ethical and unethical behavior, but there is no ironclad rule that makes the decision simple. The interviewer should view his or her behavior as though it were to be published on the front page of a Sunday edition of the local newspaper. How proud would the interviewer be if the interviewer's supervisor, parents, and closest friends read in the paper about the methods he or she used during interviewing? If the interviewer begins to rationalize or justify techniques used, then more than likely the techniques are unethical. We rationalize to make ourselves feel better about what we did. If we have to rationalize, then it stands to reason that what we did might be unethical.

Statements such as, "the end justifies the means," should be examined closely. Interviewing is a prime area in which the means are as important as the end. If unethical means are used, then the end cannot be trusted to provide valid information. Information must be provided freely and voluntarily without compelling influence (*Miranda v Arizona*, 1966).

Following the due-process model, interviewers operate under the principles that the process of the law is more important than a conviction or catching a client. Under the due-process model, it is recognized that the coercive authority of the criminal justice system sometimes abuses, and must be balanced with, an individual's rights (Peak, 2009). Even when students are not interested in becoming police officers, they should be well-versed in the U.S. Constitutional amendments and related case law. Individual rights should be protected by all criminal justice professionals.

EXERCISE 2.1

Importance of Ethics

Activity: The instructor facilitates a discussion with the students about the fluidity of ethics depending on their criminal justice role and the purpose of their interviews.

Purpose: To facilitate the understanding of the many different faces of ethical behavior.

1. What would be the difference between a police officer interviewing a suspect in which the consequences are loss of freedom and a professional interested in treating a court-deferred substance abuser? Does duty play a different role for the officer than for the treatment professional?
2. What are some of the ethical issues that a probation officer working with a client faces versus a juvenile court counselor working with a juvenile?

Ethical Standards

We described the principles of interviewing in Chapter 1. Ethical standards closely follow these principles. The interviewer begins with the understanding of the worth and dignity of every human being. Openness and nonjudgment must be maintained throughout the interview, treating each interviewee with fairness and impartiality irrespective of socioeconomic status, associates, race, religion, or physical appearance. On the other side of the same coin, interviews, decisions, and records should remain untainted by personal, financial, political, or any other type of improper influence.

Interviews are built on information gathered by all possible sources. The interviewer strives to obtain and verify additional information by methods and techniques that have the greatest probability of eliciting reliable and valid facts. It is becoming more evident that information obtained as a result of coercive techniques is inherently unreliable (Sear & Williamson, 1999). There is never a justifiable reason to suppress or falsify relevant information obtained in an interview.

Interviewing is hard work and should be established on thorough preparation. No excuse, including fatigue or disinterest, ethically permits taking shortcuts in an interview.

The English police have moved their investigative procedures away from the U.S. practices of interrogating the suspect to more thorough interviewing of victims and witnesses. The English police are trained to search for the truth rather than to push for confessions. More thorough interviewing of victims and witnesses is likely to turn up better evidence, which in turn increases the likelihood of gaining valid information about the suspect (Bull & Milne, 2004). The English judicial system also examines closely whether or not the police behave fairly toward vulnerable suspects such as those with special needs or low intelligence (Sear & Williamson, 1999). This subject will be discussed in greater detail later in this chapter.

Deception in many cases is a prime example of unethical rather than illegal behavior. There are a number of court decisions in which police officers have used deception to extract a confession, although obvious deceptive actions such as false promises that mislead a suspect and lead to confessions (*Pyles v State*, 947 S.W.2d 754, as cited in Klotter, Walker & Hemmens, 2005) are considered by the courts as involuntary confessions and are not admissible (*State v Burdette*, 611 N.W.2d 615, 2000, as cited in Klotter et al., 2005).

State v Jackson (1983) provides an ethical rather than a legal question. In this case, the police interviewed Jackson, the suspect, multiple times. Jackson was not in custody, but rather requested to talk to the police. The police believed they had found the knife used in the murder. A knife matching the set from the victim's kitchen was found thrown in the bushes near the crime scene. There was no blood on the knife and no fingerprints. One of the detectives had an idea. He took an identical knife from the set still in the kitchen. He cut his thumb, drawing blood, and then pressed his bloody thumb to the recreated murder weapon.

During the next interview, Jackson again denied killing the victim. The detective jammed the substitute bloodied knife in front of Jackson's nose and yelled, "Then how do you explain your fingerprints on the murder weapon?" The detective also told Jackson that a witness had seen him run from the victim's apartment. That was a lie. The witness saw a man running down the street, not out of the apartment, and could not identify him. Jackson began to recount what his role in the victim's murder was.

Jackson was tried and convicted. The case went up on appeal.

The appellate court's decision ruled that Jackson's confession was admissible, despite the detective's deception. The court's decision, however, included the reprimand ". . . the officers deceived and lied to the defendant. Such actions are not to be condoned by the courts . . ." (p. 21). So while the detectives' deceptive actions were legal, their ethical behavior was questionable.

Deception is on the "slippery slope" toward unethical behavior. As an interviewer, detective, private investigator, or social worker begin to slide, it becomes more and more difficult to define what behavior is acceptable and what is not.

Validity of Confession Evidence

Although this book is restricted to interviewing and deals with interrogations in a limited fashion, many of the ethical issues surround pressuring individuals to confess to acts they may or may not have committed.

Folklore states that an innocent person would never confess to a crime that he or she has not committed. Because juries believe this folklore, confessions are one of the most powerful pieces of evidence that can be admitted into court (Gudjonsson, 1993). There is a growing body of literature, however, that has examined the vulnerabilities of people and the risk of false confessions. Innocent people, in fact, sometimes do confess to crimes they have not committed.

Although the incidence of false confessions is unknown, Leo (1996) discovered a disturbing number of documented cases in which defendants confessed, later retracted the confession, were convicted at trial, and then later exonerated by DNA or other forms of irrefutable physical evidence.

Researchers for The Innocence Project also have found that roughly a quarter of all DNA exoneration cases contained full or partial confessions, apparently false (Leo, 1996). Probably one of the most famous cases was the Central Park Jogger. In 1989, Trisha Meili, left for dead after being raped and beaten in New York's Central Park, had no memory of the attack against her. After confessing to the police, five teenagers were convicted of her attack and sentenced to 7 to 11 years in prison. Nineteen years later, the convictions were thrown out after a convicted serial rapist and murderer, Matias Reyes, told prison officers that he alone attacked Meili. DNA testing confirmed his statement. Five young, innocent boys spent much of their lives in prison due to their vulnerabilities that led to false confessions.

Dr. S. M. Kassin (1997) describes three types of false confessions. They are voluntary, coerced-compliant, and coerced-internalized.

A voluntary false confession is offered without external pressure from the police. Quite often people will confess to crimes that are highly publicized. They desire fame, recognition, or self-punishment. For instance, when Charles Lindbergh's baby was kidnapped, 200 people went to the police and confessed. Because of Charles Lindbergh's famed nonstop flight from New York to Paris in the single-engine monoplane, Spirit of St. Louis, these 200 people had a distorted fantasy that by confessing they would achieve fame.

None of them had kidnapped or murdered baby Charles Lindbergh, Jr. (FBI History, n.d.)

On the other hand, coerced-compliant false confessions occur for instrumental purposes. Individuals in these circumstances confess to escape or to avoid harsh interrogation. They are made to believe that the short-term benefits of confessing outweigh the long-term costs. This is particularly true if their perception of the strength of the evidence is quite high.

Individuals who are predisposed to exhibit compliance may be particularly vulnerable. For example, Dr. Kassin (1997) describes a case in Great Britain in which a 17-year-old was arrested for murder despite the absence of any physical evidence implicating the youth. After being interrogated for 14 hours by five detectives, the boy was exhausted and felt powerless to bring the interrogation to an end. He confessed. After a night's rest, just one day later, he retracted his confession.

Even more worrisome to the criminal justice community are cases of coerced-internalized false confessions in which the individuals actually believe they committed crimes. Individuals who can be coerced to the point of believing they committed the crime tend to possess poor memories, high levels of anxiety, low self-esteem, and lack of assertiveness. Their suggestibility scores increase with sleep deprivation.

Dr. Kassin (1997) describes Peter Reilly, an 18-year-old, who found his mother dead and called the police. During the interrogation, Peter was polygraphed and then told that he failed the polygraph test. Transcripts of the interrogation reveal that Reilly went from denial, to confusion, to self-doubt, and finally, to confession. Peter signed a full written confession. Two years later, independent evidence revealed that Reilly could not have committed the murder.

Thomas Sawyer was accused of raping and murdering a neighbor. Sawyer was interrogated for 16 hours. During the interrogation, Sawyer was told that his hair was found on the woman's body, although that was not the case. Sawyer moved from adamantly denying the accusations to believing he committed the crime, but lost his memory as the result of an alcoholic blackout. Sawyer even altered his story to match the crime scene. After denying for hours, Sawyer confessed, concluding that, with all of the evidence pointing toward him (fabricated), he must have killed his neighbor (Kassin, 1997).

The primary factors that often lead to individuals confessing to crimes they did not commit are their vulnerability and the presentation of false evidence (rigged polygraph or forensic tests). Vulnerability factors are youth, lack of experience with the criminal justice system, lack of intelligence needed to comprehend the interviewer's techniques or the consequences of a conviction for the offense, fatigue, isolation, and intoxication from alcohol or drugs.

While falsely telling a suspect that he or she failed a polygraph was at one time common, what is becoming more common is the police telling a suspect that the analysis of physical evidence places him or her at the crime scene, or that the weapon used has been found and incriminates him or her.

Eye witnesses then often "appear," claiming to have seen the defendant at the crime scene.

Dr. Kassin (1997) conducted a lab experiment using college students as subjects. Kassin wanted to discover if presentation of false evidence leads individuals, who are vulnerable, to confess to acts they did not commit, to internalize (or believe) their confession, and to make up details to support their false confession.

During the experiment, students were instructed to type letters as quickly as possible. "Don't press the ALT key," the researcher told them. After 60 seconds, the students' computers supposedly crashed. The students were accused of hitting the ALT key.

After the students were prodded to sign a confession stating they hit the ALT key, they met another associate outside the lab door who asked, "What happened?" After the students told the associate, they were brought back into the lab and asked to reconstruct when and how they hit the ALT key. Vulnerability was manipulated by varying the speed of the typing requirement and the presentation of false incriminating evidence. The associate falsely told half of the students that he or she saw the students hit the ALT key.

Overall, 69% of the students signed a statement confessing to hitting the ALT key; of these, 28% internalized the guilt, and 9% created details to match their confession.

When students were allowed to type at a slower pace, and there was no associate who stated he or she had seen the act, 35% of the students signed a confession, but none internalized or made up information to support the confession. When the students were pressured to type quickly, all of them signed a confession; 65% believed they had hit the key, and of those students, 35% made up information.

The laboratory experiment was conducted on individuals who did not meet the criteria of being vulnerable. They were college students who were not using drugs or alcohol or facing the possibility of losing their freedom. Still, given stressful conditions and false accusations, they confessed to acts they did not commit. Many of the students actually believed they had hit the key and confabulated details. How much easier it is, then, for people in authority such as police officers or probation officers to convince vulnerable people that they committed criminal acts.

Dr. Kassin followed his initial study with an examination of individuals who waive their rights (2005). In *Miranda v Arizona* (1966), the U.S. Supreme Court ruled that the police must inform suspects in custody of their Constitutional right to remain silent and to have an attorney.

Dr. Kassin concluded that individuals who do not have prior felony records, juveniles, and innocent people are most likely to waive their Miranda rights and to talk to police officers without an attorney present. On the other hand, Dr. Kassin found that suspects who are actually guilty believe if they insist on their rights to silence and an attorney, the police will think they are guilty.

Innocent suspects waive their rights because they are innocent and have naïve faith that their true thoughts, emotions, and other inner symptoms can be seen by others. So, even with the protection of their rights to silence and to an attorney, many people elect to talk to the police, increasing the possibility of false confessions. The officers may present what appears to be such overwhelming evidence that they believe the individual is guilty and that the individual is thrust into a state of despair. Then, the officers minimize the seriousness or the consequences of the crime, provide moral justification or opportunities for face-saving, and make confession seem like an advantageous means of escape. Although the officers do not make any promises to the individual, they state they will tell the prosecutor that the individual was cooperative.

Interviewers should always remember that confessions can only be considered valid if they are given freely and voluntarily, if they are not tainted by an illegal arrest or search, if the interviewees are provided counsel if they desire it, and if they are given Miranda warnings if held in custodial conditions.

EXERCISE 2.2

Applied Ethics

Activity: Consider inviting professionals to share their real-life ethical dilemmas.

Purpose: To introduce students to the realistic struggles of remaining ethical in difficult situations.

Explain to the students the professional's job. Before the guest professional comes to class, brainstorm with the students some of the ethical issues surrounding interviewing they believe the professional faces. What are some of the questions the students would like to ask the professional? It is recommended that the questions be sent to the professional before the class presentation so he or she has time to think about them. Ethical issues are threatening, and it takes a stable, well-centered person to be willing to discuss issues he or she has faced.

Ethical Tools

Entire courses are taught on ethics in criminal justice, and criminal justice professionals should be encouraged to participate in them. These courses help professionals, and more specifically interviewers, think through independent moral decisions based on sound logic and good reasoning and not just what feels right or the opinions of their peers.

Robert Pring (1990) discusses the need to introduce moral logic as a series of steps:

1. Establish the goal of the moral or ethical deliberation
2. Establish the means to attain the goal
3. Establish the action, which naturally follows

So for example, (1) the detective wants to reveal the facts of a specific crime; (2) the detective knows the physical evidence, and complete, valid interviews of the victims and witnesses are required to reveal the facts of the crime; and (3) the detective must conduct a thorough, prepared interview in conjunction with a thorough discovery and analysis of the evidence.

While interviewers can have other goals (e.g., get a confession), ethically the interviewers need to consider the moral value beyond the action. For example, is a confession really the goal or is the goal conducting a thorough investigation that includes well-prepared interviews of all related parties that culminate in statements that corroborate each other and the evidence?

Criminal justice professionals also must consider duty-oriented moralities—acting exactly as duty dictates (Pring, 1990). Most criminal justice professionals are required to take an oath to uphold the law. It then becomes important for every professional to decide how strictly he or she will follow his or her oath, especially when it conflicts with other moral goals he or she may desire to achieve. For example, probation officers may have the duty to report every infraction clients make, but the officers know they would spend all of their time writing up most of their clients.

Is there a better way to reach certain goals with their clients? The question then becomes, how do criminal justice professionals reconcile their duty with other goals?

When preparing to interview, the interviewer must think about the moral goals that must be achieved and by what means. There is also the need to think about the duties of the profession and how strictly they should be followed if there are conflicts between the interviewer's moral goals and the duties of the profession.

Confidentiality

The diversity of agencies in the criminal justice system does not allow for a general statement about confidentiality that covers all situations.

Krishna Samantrai (1996) details the importance of confidentiality in health and human services. While limited in disclosures of child abuse or threats of harm, the workers have an ethical obligation to preserve confidentiality about the client's personal lives. Part of that ethical obligation in health and human services, as well as other agencies, is the need to inform the client what can and cannot be kept confidential. While there may be some incidents in which certain aspects of the client's information are shared with other professionals, it is never discussed in casual or social settings or with other clients.

The goals of law-enforcement interviews are usually crime-related rather than helping-oriented; however, wherever possible, confidentiality is still of ethical concern. Suspects in custody are detailed their rights, but victims and witnesses need to know what information will be shared with other professionals. It is unprofessional and unethical for detectives to discuss specific cases over a beer with friends.

Conclusion

Interviewers shoulder a heavy burden to be ethical because they possess great authority and power. They are, in fact, the physical representative of the city, county, state, and federal government, regardless of the agency they work for.

Each interview conducted should be completed with the goal to gather information, but not at the expense of the individual's self-esteem.

Research has shown that people can be made to confess to crimes they did not commit. Therefore, it is critical that interviewers prepare extensively so they are familiar with the individual and the case.

All criminal justice professionals, and specifically interviewers, need to decide what the moral goals of the interview, specifically, and their profession, generally, are. These goals need to be reconciled with the performance of their duty. The public must have confidence in the criminal justice professional's integrity and professional standards.

A guiding rule before acting in a questionable manner is, if the interviewer's spouse, parent, or other family member read about the action in the newspaper or heard it broadcast on radio or television, would the interviewer feel pride, or shame? No professional ever wants to open the morning paper and see his or her name connected with a scandal.

EXERCISE 2.3

Practicing Ethics

Activity: The instructor divides the class into small groups and gives each an ethical interviewing dilemma. The small group discusses all the moral dimensions of the problem and how they would deal with it. One person reports to the entire group.

Purpose: To discuss ethical issues during interviewing that the students may face when employed by the criminal justice system.

1. As a child protective service worker, you are expected to investigate cases of child abuse and neglect. You receive a complaint from a neighbor of Sally Jones who states that she hears Sally's baby crying day and night. Further questioning of the neighbor does not provide additional information, except that the neighbor says that Sally is young, unmarried, and has no business having a baby. You decide to walk next door to Sally's house unannounced. A young, tired girl comes to the door. Behind her you can see a very dirty house, but you do not hear the baby. Without considering legal implications,

what are some of the ethical issues that you need to consider at this point? What solution will serve the social and moral good?

2. You are a probation officer who has an overwhelming caseload. You have one client who always shows up late for appointments, has not been able to find a job, and smells suspiciously like alcohol. He has a few probation violations, but none sufficient to be sent to prison. You know your job is to try to help your clients become productive citizens, but you are beginning to think if this client does not care, why should you? What are some of the ethical issues you should consider before this client comes in (probably late) for his appointment with you? What do you do? What solution(s) will serve the social and public moral good?

3. You and your fellow police officer are patrolling a neighborhood known to have a high level of drug traffic. You see a group of young boys hanging out in an empty lot. Your partner says, "Hey, Paulo is part of that group over there. I bet if we call him over and start acting real friendly and maybe hand him a few dollars, he will get burned big time by his buddies." What are some of the ethical issues you should consider? What is the moral goal? What are the means to get to the goal? What do you do?

4. You are the detective investigating the murder of a young boy. You are sure that John Brown is the perpetrator, but no physical evidence has been found to connect him to the crime scene or to the body. You decide to tell John Brown that somebody saw him with the young boy the evening of the murder. What are some of the ethical issues you should consider? Are there specific social and public goals to consider? What is the overriding moral goal?

5. You are partnered with a police detective who is interrogating a man for sexual assault. The man states that he met the victim at a night club. After dancing and drinking together, he gave her a ride home. He states that he dropped her off at her door, and that was the last he saw of her until he was arrested for sexual assault. The detective says to the suspect, "Look, John. I know how it is. This nice looking woman spends all evening with you, rubbing up against you, and letting you buy her drinks. And then, she gets you to take her home. What does she expect? What a tease! I don't blame you at all. I would have done the same. You can tell me. She invited you in, but then after consenting, she got scared that her boyfriend would find out so she called it rape! Isn't that the way it happened, John?"

What ethical issues has this detective caused? What do you think you should do? Would it make any difference if you were certain this man was the suspect? What are the moral goals? What means should you use to get to the goal? What actions should follow?

6. During the interview with one of your key witnesses, you realize that the time frame of what she is telling you does not support your case against the suspect. When you ask questions to clarify, she becomes confused. You consider whether you should include the witness' information in your report. What ethical issues should you consider? What do you do?
7. While still a student, you intern with the sexual assault unit of your local police department. Another student is interning also. In the intern class, the other intern begins discussing what a victim said. Because the case had made the news, it is clear who the intern is talking about. The instructor does not stop the intern. What ethical issues should you consider? What should you do?

ALLEN'S WORLD

The North Carolina Association of Private Investigators has some of the following rules in its Code of Ethics:

> A private investigator is dedicated to a search for truth and the furtherance of his client's interest consistent therewith. This search for truth makes possible the establishment of the American ideals of fairness and justice for the benefit of the client in every case that the investigator works on. It should be the intention of every investigator to deal honestly, justly, and courteously with all persons and to practice his profession according to this Code of Ethics.
>
> HE/SHE will make all his reporting based upon truth and fact. He/She will not disclose or relate or betray in any fashion that trust of confidence placed in him by either client, employer, or associate, without that person's consent. He/She will not suggest, condone, or participate in any fashion or degree, for any purpose whatsoever, in entrapment.
>
> HE/SHE will not compete illegally with other investigators in the solicitation of work and not engage in the unauthorized practice of law (North Carolina Association of Private Investigators, n.d.).

As for me, I am concerned that my clients know what I can and cannot do for them. I believe ethically that potential clients should know what I will need to do to meet their requests. Then they ask me what it will cost. I estimate the price, but I also tell them what variables will play into the expenses and fees. I estimate maximum costs. I think it is wrong to reel them in by estimating low and then piling on junk fees.

I almost always take a retainer. If I do not use all of that money, I return the difference. Many private investigators (PIs) keep the entire retainer regardless of the work and/or expenses involved.

I often assist adopted children in finding their birth parents. Up front, I tell them the chances are 50–50. And, I reduce my fee for these cases as a way of making me feel that I am more than just a paid employee.

I meet with potential clients to discuss cases. I do not charge for these sessions if I do not take the case.

If I am working directly for the client, I provide continual updates and make recommendations as to when the work should stop. I want to spend my client's money as if it were mine. And by nature, I am a frugal person.

I encourage them to talk to an attorney before they hire me and possibly waste money. Why? Many clients who call me directly are desperate. They have talked to well-intentioned friends who gave them sage advice. Their friends' advice most often happens to be wrong and of no legal value. I want potential clients to hear directly from an attorney what will and will not help them legally. I do not want to waste my clients' resources.

Let the attorney explain the facts of life and, if possible, suggest an investigative route that might be legally beneficial.

For example, a client calls and asks me to find somebody. That person owes the client money. The client expects that when I find that scofflaw, he or she can take that debtor to court and collect the money owed the client. I can find that person, but the client is not likely to get the money owed to him or her. That is because in North Carolina, it is almost impossible to enforce a judgment. I often talk these clients out of throwing good money after bad.

I will not change the truth I find in my investigations. Attorney Tom Harrow, now deceased, once asked me to investigate an accident involving his client. The young man had been run over and was a paraplegic, confined to a wheelchair for eternity.

Harrow saw $$$$$ signs—big $$$$$$ signs.

My investigation showed that the young man was drunk. His blood-alcohol content exceeded the legal limit. He had passed out about 2 A.M. on a hump in the road. It was dark and raining, and he was wearing dark clothes. A car came over the hump and hit the young man.

I got a call from Harrow when he read my report. He wanted me to leave out a few facts. Like, did I have to say the victim was passed out, drunk, in the middle of the road?

Did I have to say it was raining, there were no street lights, and he was wearing dark clothes?

I told Harrow I could not, would not, change my report. Harrow did not take the case to trial. He never used me again.

Sad to say, there are PIs I know that willingly change reports to maintain the goodwill of the attorneys they work for—and to keep the cash register ringing. Private investigators are not bound by the same restrictions as the police. I do what my conscience allows me to do, my internal moral code, in getting information that helps my client. What are my restraints? I have to worry how my methods will look in court if I have to testify or that the attorney or client I am working for might be offended.

One of my golden rules of ethics is that if someone has broken a law to gain an advantage against my client, then I have no legal or ethical restraints in getting the information I need. That is a rationalization that makes me feel good about what I do and allows me to represent my client most effectively. My rationalizing can easily be labeled utilitarian ethics or situational ethics by critics. The ends justify the means for me if I believe I am not hurting an innocent person. The key phrase is "innocent person."

When a financial officer embezzles money from the company I am representing, I feel justified in searching his or her financial records to see where the money is hidden. That embezzler is no longer an innocent person.

When trying to figure out if a husband is cheating on his wife, I feel justified in getting a record of his long distance calls and his cell phone calls to see who he has been calling.

I often tape record phone conversations and in-person conversations without telling the subject.

I can secretly get financial information. I will go through garbage, if it has been put out on the curb, public domain, looking for evidence.

I might wrongly tell a witness that another witness has contradicted his or her statements.

In one case I investigated, an employee stole a patented formula from his company for making marking crayons that could be erased without leaving a mark. He established a business to produce marking crayons and undercut his former employer. I went through his garbage and found molds for the crayons. I found prototype labels.

Not good enough, said the attorney who hired me. What do I do now?

I had a friend who worked in a tax office. I went to see my friend and asked if I could use his phone. I called the man who stole the patent.

"Is this Robert?" I asked. He said yes.

"My name is Allen Cowan, and I'm calling you from the Gaston County tax collector's office."

I told Robert he had to fill out forms so the tax assessor would know how to tax his business. I told Robert I would bring the forms to his office, but I needed to ask him some questions so I would know what forms to bring.

"Is your business wholesale or retail?

"How many employees?

"What type of equipment?

And then, the only question I was interested in, "What is your product?"

Robert told me he was making marking crayons. My work was done. The attorney I worked for filed an injunction. (An *injunction* is a court order telling someone to stop doing what they are doing. In this case, stop making marking crayons.) Robert was shut down. He never opened his business.

Ethical? If you read my words carefully, I never lied to Robert. I misled him, but I did not lie.

Here are some real phone calls I got that involve ethical issues. Do not laugh. I really got these calls. How would you, the reader, respond?

A potential client calls. He wants to hire me to follow him and then write a report that says he is not cheating on his wife. He tells me he will drive slowly, will tell me where he is going, and will let me take videos and still pictures. Would you take the case?

A potential client calls. Two days earlier, on Valentine's Day, his wife was driving his car. The cell phone rang, and a woman's voice said, "Thanks for the flowers. That was so sweet of you." The wife asked who was calling. When the female caller heard a female voice on the other end, the caller hung up. The potential client wanted me to trace the incoming cell phone call and tell his wife the caller had the wrong number. Do you take this case? Why or why not?

A potential client calls. Her house is haunted. "Can you come over and exorcise the ghosts?" Do you take this case? Why or why not?

I got a call from a pimp in Gastonia, North Carolina. One of his girls was hooking on the side and not paying him his "finder's fee." He wanted me to sleep with the woman, give him the time and date, what sexual act she performed, and how much I paid. He would then check to see if she gave him any money. Do you take this case? Why? Why not?

A potential client called. She was engaged to a man. She suspects he is fooling around. He goes to a certain bar every Friday night. She wanted me to send in a female undercover operative, play coy, and see if her fiancée made any advances. Do you take this case? Why? Why not?

Here's the answer to the first example.

I did not take the case. The potential client wanted to use me to convince his wife he was being faithful. To me, that is unethical. You decide if you would take this case, keeping in mind you are losing an easy fee and that there are a lot of PIs out there who would, and one of them probably did, take this case.

As a reporter, I once snuck into a private meeting of developers who were going bankrupt. They did not want the public to find out. Nobody asked me who I was, or what I was doing there. I just walked in and sat down. The moderator said he knew reporters were in the audience. He asked any of us to leave. What did I do? I stayed and took notes secretly. When the meeting ended, I introduced myself to the moderator and said I had a few questions.

"You can talk to me," I said, "or I can print the article without any clarification." He talked.

When I got back to the newspaper office, I told my editor what transpired. Did I do the right thing by staying? "Hell yes," said Stan Brennan (now deceased). The next day's paper had a front-page article about the meeting and the bankruptcy speculation.

Here's my all-time favorite story involving ethics some might question.

Paul McDonald, an investigative news reporter who worked with me, had a PhD. We called him "Doc." One night, there had been a train wreck at the crossing near Atando Avenue. Victims were taken to three separate hospitals. Two of them had public relations people on hand to answer media inquiries. The third, however, had closed shop for the night.

Paul McDonald called the emergency room, and the call went something like this:

"This is Doctor McDonald calling. I'd like some information about victims of the train wreck." Remember, McDonald had a PhD and did have the title of doctor before his name.

The folks in the emergency room answered all of his questions. The next day's newspaper for which Doc worked had the only complete information about the number of victims and the extent of their injuries.

When I got to the office the next morning, I saw Paul in the editor's office amidst a group of official-looking, very serious people. When Paul came out of the office, I went over and asked him what was going on.

"Those people were from the hospital," he said. "They were complaining that I lied to them."

The editor reprimanded Doc in front of the hospital officials. Later, the editor called Doc back into the office and congratulated him.

Paul did not lie to the folks in the emergency room. They blamed him for getting information under false pretenses. When in fact, all they had to do to uncover the deception was ask a simple question, "Do you have privileges to practice here?" You, the reader, can do the mental gymnastics. Paul McDonald: Ethical? Or corrupt?

Older readers of this book might remember the *Chicago Sun-Times* exposé back in 1978.

At the time, building inspectors in Chicago were notorious for taking bribes. The *Sun-Times* bought an old tavern, went about rehabilitating it, and then actually opened it for business. They called it, fittingly, The Mirage. Staffers included *Sun-Times* reporters and photographers.

Beginning in late 1977, while the rehab was underway, many of the inspectors came in to inspect, and to solicit bribes to look the other way. When the tavern opened, the *Sun-Times* ran it for 4 months, gathering more evidence against the inspectors.

On January 8, 1978, the *Sun-Times* began a 25-day, front page series on the bribery and rampant corruption. Charges were filed. Inspectors went to jail.

At the time, I was on a fellowship at The University of Michigan studying law. There were 12 of us reporters in that National Endowment for Humanities group (a federally-funded program that paid selected reporters to go to school for a year). All 12 of the fellows agreed what a great series it was. Surely the *Sun-Times* would receive a Pulitzer for public-service reporting. How wrong we were. Ben Bradlee of the *Washington Post* and Eugene Patterson, editor at the *St. Petersburg Times,* lobbied against the award for the *Sun-Times.* They both said that any series based on deception should not be rewarded for the deception. The Pulitzer committee voted against honoring the *Sun-Times,* saying "truth-telling enterprises should not engage in such tactics" (Lisheron, 2007).

Again, you, the reader, can do the mental gymnastics. The *Chicago-Sun Times:* Ethical? Or corrupt?

Preparing to Interview

OBJECTIVES Upon completion of this chapter, students should be able to

1. Develop objectives for an interview
2. Develop relevant questions to meet the objectives of an interview
3. Develop motivating questions that will encourage the interviewee to respond, enhance memory, and get more informative responses

Introduction

In the movies, the detective grabs the bad guy and begins firing perfectly developed questions that result in relevant answers and a confession. Within the 2-hour movie, there is little indication of the preparation that must take place to conduct a relevant and complete interview.

Effective interviews must have clear objectives—what does the interviewer plan to achieve by the end of the interview? Without objectives, the interviewer cannot assess whether the interviewee's responses were relevant, informative, and complete. It then becomes unclear if the interviewer needs to ask follow-up probes to gain more complete and clear answers.

The questions must be relevant, developed to meet the objectives of the interview. The questions need to be motivating, designed in such a way that the interviewee is willing to respond with valid and complete answers. Even the most motivated interviewees need questions structured to help enhance memory retrieval. It takes skill, training, and patience to help witnesses remember details.

> The probation officer closed the file on the new client. After reading the information compiled for court, the original police reports, and the additional information from the defense attorney, the officer had a pretty good idea of the client's background, criminal record, work history, and drug use. Although the officer would continue to collect information during the probation period, the first meeting with the client was that day.
>
> The probation officer thinks, "The objective of the first meeting is to assess the client's understanding and feelings about the terms of probation. Let me think what the best questions would be to obtain hopefully honest answers."

Interview Objectives

It is rare in the criminal justice field to have established objectives for interviews that are to be conducted. With the exception of "booking" interviews that newly arrested offenders encounter when processed in jail, there are few interview situations in which criminal justice personnel just complete forms. More frequently, each client, citizen, victim, witness, or situation is unique and clear objectives for each interview are required. No two interviews are identical.

In the example of the probation officer, a "shotgun" approach in which the probation officer begins to fire questions from all directions is counterproductive and might alienate the probationer.

Also, as we will describe in Chapter 4 in the discussion about interview phases, during the beginning phase of each helping interview the objectives should be clearly laid out. So if the first meeting's objectives are to assess the client's understanding and feelings about the terms of probation, then the probation officer needs to formulate questions around the probationer's understanding of the terms of his or her probation and what factors will facilitate or hinder the client's ability to keep the terms. The officer also will gauge the probationer's feelings about probation, such as feeling overwhelmed or resentment due to the conditions, guilt, or shame toward himself or herself, and antipathy toward the officer.

The probation officer realizes that he or she will be seeing this new probationer often during the following weeks. It is critical that objectives for each interview are established before the probationer arrives so the officer can assess what has been achieved and what still needs to be accomplished.

For example, the objectives for the first interview were the client's understanding and feelings toward the conditions of probation. The objective of the second interview may be the probationer's adaptation to the conditions of probation. This specific objective helps keep the interview focused on the probationer's ability to cope with and adapt to a new way of life. It would be easy to overwhelm the probationer with a battery of questions dealing with all of the conditions. Has he or she found a job? Has he or she been staying away from associates who are offenders? Has he or she been following the curfew? All questions for which the probationer will formulate excuses and begin to feel more and more defensive. The probationer could easily leave the interview feeling misunderstood and unsupported.

A more supportive approach to the interview might begin, "John, I would like to hear how you are adapting to the conditions that have been placed on you. I can imagine that it has taken some adjustment." This form of open-ended, but specific, declaration of the interview's objective welcomes the probationer to problem solve with the probation officer, but at the same time, the question is specific.

Below are some examples of how the probation officer can keep the probationer on track as necessary by using the interview's objective as a reminder. The probation officer can assess the success of the interview by how well the objective was met.

"John, from what you have told me you have excellent family support. Your wife and parents are helping you stay away from past associates and keeping curfew."

"Your greatest difficulty is finding a job because your driver's license was revoked. We discussed other alternatives for transportation and the possibilities of petitioning the court for limited driving privileges."

"John, I feel very positive about your chances of success. Even though you were frustrated when you came in about the loss of your license, you do understand the reasons and are realistic about the need to show the court that you are becoming trustworthy. Next week, we will see how the transportation and job opportunities are coming."

In another example, a community police officer wants to interview a number of citizens about their perception of crime in their neighborhood. The officer's ultimate objective is to assess their willingness to organize a community watch. So, what exactly is the objective or objectives of the citizen interview?

The questions constructed will differ greatly depending on the ultimate objectives. If the officer wants to gather information about citizens' perception of the crime problem in their neighborhood, he or she includes questions about the types of crimes the citizens have experienced, the scope of the crime, the impact of the crimes on the residents, and perhaps what they see as the underlying problems and/or solutions to the problem. On the other hand, if the officer has concluded that the neighborhood could benefit from a community watch, the officer will assess the commitment he or she might get from the citizens. The questions will be directed more to the residents' willingness to invest the time and effort needed, although the residents' general concern about crime and their perceptions of the effectiveness of a community watch are important.

In either case, there are a number of concepts that must be defined precisely. For example, what are the boundaries of the neighborhood? Is it a well-defined subdivision or a subsidized housing area? Is it a gated community? Does it have a guard gate where visitors must register? Does the area include any business areas, or is it completely residential? Are there regular police patrols? Are there streetlights? It must have clear boundaries.

Crimes must be defined. Are they to include personal and property crimes, vandalism, or underage drinking? A time limit should be included. Most communities are dynamic and change over time, so it will be important to delineate the frequency of crime in the past year, the past 5 years, or longer.

Because citizens' perceptions will include those of their neighbors, extended family, and business associates, it will be important also to demarcate what will be considered a family.

An example of the objective might be: To explore residents' perceptions of criminal activity within the Cherry Ridge Community over the past year by obtaining data about property and personal crimes they or their immediate family members living in the community have experienced.

The above objective must be concise, particularly because it will be used to develop a set of questions asked of many people. The same precision is needed when interviewing each individual.

EXERCISE 3.1

Writing Objectives

Activity: The students will identify the criminal justice professional with whom they will conduct their 15-minute interview and write the objectives for the interview.

Purpose: To practice writing objectives and receive feedback from other students and the instructor.

The students should be instructed early in the semester to think about a criminal justice professional they would like to interview (see Chapter 14 for more detail). They should be instructed not to pick somebody they know, but rather a representative from a career they might be interested in pursuing. The instructor should be prepared to help them make the necessary contacts.

In class, students present the agency in which they are interested and why. The students should not be critiqued, although the instructor can place some limits on the students' selections if the possibility of obtaining interviews with representatives of particular agencies are not realistic, e.g. CIA.

After the discussion about objectives, students are to write their objectives for their interviews. These objectives should be shared in small groups with discussions about any unclear terms and realistic achievements for the 15-minute time frame. The instructor also should examine each objective and provide necessary feedback. These objectives are the foundations for their questions.

Formulating Basic Questions

Once the objectives for the interview are established, the questions should relate to those objectives. It sounds simple, but it takes skill and practice before it is easily accomplished. The interviewer must keep in mind the information that is needed to achieve the objectives but, at the same time, must develop specific, concrete questions.

For example, if the probation officer asks the general question, "How do you feel about the terms of your probation?" the officer knows the question is relevant. The issue then becomes whether or not the probationer receives the question the way it was intended, and does not misinterpret the question and wander in a variety of directions.

Later, discussion about the use of open- and closed-ended questions will describe the advantages of open-ended questions. No matter the type of question, it still must provide direction.

The general question should be broken into simple questions that move the interviewee in the direction that will provide a complete answer to the entire question. In developing simple questions, the interviewer must be sure he or she is targeting the correct dynamic. The probation officer may be thinking broadly about the probationer's feelings about the probation terms, but the concern may be more on the probationer's capabilities to fulfill those terms, which is not as much about feelings as it is about actions.

Also, feelings about something out of the probationer's control may lead the interview down a path that is not constructive, especially at the beginning of the interview. So, individual questions may be about each of the conditions of probation and the probationer's capabilities of fulfilling them.

For example, if the probationer is expected to meet with the probation officer weekly, the probation officer might ask, "Your first condition of probation is to meet with me here in this office each Thursday at 2 P.M. Is there anything that might prevent you from making these meetings?"

Toward the end of the interview, the probation officer might wish to return to the probationer's attitude toward the probation conditions because it will provide an overall sense of how likely the probationer is to succeed.

EXERCISE 3.2

Identifying Relevant, Concrete Questions

Activity: Given the objective of an interview and a list of questions, the students will decide what questions are relevant and sufficiently concrete.

Purpose: To practice recognizing effective basic questions before writing their own.

The purpose of the interview is to discuss Paul's truancy with his mother. Although his mother is responsible for Paul's behavior, he is in middle school and is supposed to take the bus to and from school. The truancy officer wants to know if Paul's mother is aware that he has missed a great deal of school and what her culpability is.

Are the following questions adequate to achieve the objectives? If not, write additional questions that would be needed. Delete any questions that do not appear relevant.

Questions:
What transportation do you use to get to work?
What is your work schedule?
Who else lives in your house besides you and Paul?
How is Paul doing in school? Does he like it?
Does Paul complete his homework?
How does Paul get up in the morning?
Does he eat breakfast?
Do you have any communication with Paul's teachers?

Words   As the interviewer is developing questions, it is vital to consider the interviewee's vocabulary comprehension. Such factors as the interviewee's level of education and area of employment may provide clues to vocabulary. To be safe, use simple, direct language. Avoid technical terms, abbreviations, acronyms, and professional slang.

The interviewer should provide working definitions of professional terms that cannot be avoided. If the agency's procedures or laws must be included, the interviewer should be prepared to provide explanations and definitions.

Even if the interviewer believes that the interviewee will be most comfortable with street language, slang, or hand gestures, the interviewer needs to be careful about venturing into that lexicon. There is the risk that the interviewer will be perceived as superficial and not to be taken seriously. The interviewer should be certain of the meanings of street jargon and observe the interviewee carefully when beginning the use of idioms.

The interviewer also should avoid words that might be unintentionally, emotionally loaded, such as *kill*, *steal*, *lie*, and *rape*. Even victims of sexual assault often are uncomfortable with the use of the term *rape* to describe what happened to them. The first year that the National Crime Victimization Survey (Bureau of Justice Statistics, 2004) was conducted, researchers quickly learned that asking respondents if they had ever been victims of personal crimes such as rape would result in a quick no. If the researchers asked a number of specific, but neutral, questions, they received more honest and positive responses.

For example, the following is an item from the revised National Crime Victimization Survey:

> 43a. Incidents involving forced or unwanted sexual acts are often difficult to talk about. (Other than any incidents already mentioned) have you been forced or coerced to engage in unwanted sexual activity by
> a.   Someone you didn't know before
> b.   A casual acquaintance
> c.   Someone you know well? (Bureau of Justice Statistics, 2004)

Whenever potentially loaded words cannot be avoided, then the interviewer should explain why they need to be used.

> "I may use terms or words that you will find uncomfortable. However, it is important to document exactly what happened. Take your time. If there is a question that makes you feel uncomfortable, and you want to write your answer, feel free to do so."

Formulating   As noted in the Introduction, an interview is a directed conversation.
Questions That   Interviewees have knowledge or experience that interviewers want to know.
Motivate Responses   The expectation is that the interviewees choose to give complete and accurate responses.

In preparing for the interview, the interviewer needs to anticipate the interviewee's level of cooperation and ability to give accurate information. Beginning with the most obvious, the interviewer needs to consider the role of the interviewee. Cooperation is likely to differ if the interviewee is a victim rather than a witness or, certainly, a suspect.

As described in the Introduction, the motivation of the interviewee is important. The less motivated the interviewee is, the more the interviewer must consider the means to motivate. Victims, in general, are more motivated than bystanders who just saw the crime, although the interviewer should be aware that victims may not want to give complete responses for a number of reasons.

Probationers may become more motivated when they realize their probation officers have the authority to improve their conditions of probation, increase the level of probation, or send them to prison.

For each of these interviewees, the interviewer must identify and plan the route to encourage valid, complete, and relevant responses. Even neutral bystanders can be motivated by the interviewer who, at the beginning of the interview, introduces the unique qualifications that the bystanders possess to help the investigation along and also expresses appreciation for their time. An example of an introductory statement might be: "I really appreciate your willingness to remain at the scene as a responsible citizen. I know you were late getting to your appointment. We need your objective observations to understand clearly what happened."

Raymond Gorden (1998) notes that if people are pressured to talk before they are willing to respond, talking does not stop, but inaccurate information begins. If unwilling, they may use a number of alternatives. They may lie or withhold certain portions of the information to avoid embarrassment or other negative consequences. They may attempt to distract the interviewer and lead the interview down an irrelevant path. They may decide they are not going to be cooperative and refuse an interview.

Gorden recommends that interviewers minimize "ego threats" and maximize empathy. The interviewee must feel that his or her sense of self is protected. Empathy will be discussed in greater detail in Chapter 5. Simply put for now, the interviewer must convey understanding.

If individuals feel their sense of self-worth, or ego, is being threatened, they may avoid answering questions, attempt to respond with answers that will keep their self-esteem intact, or provide information they believe the interviewer wants to hear. Neutral questions that do not include judging statements nor provide clues to what the interviewer may want to hear protect the interviewee's self-esteem.

If, for example, the interviewer is speaking to parents about their underage child who has been caught buying alcohol, the interviewer should ask questions in such a manner so that their parenting is not judged. If the parents feel a need to protect their self-perceptions, egos, or reputations as responsible parents, they will become defensive. Their responses will reflect their defensiveness by deflecting and protecting. For example, the question, "Were you aware that your child was drinking alcohol?' implies that responsible parents would have or should have known and prevented their child's drinking. On the other hand, the question, "I know how hard it is to keep track of an active adolescent. Do you have an idea when your son began drinking alcohol?" considers the parents as problem-solving partners with the interviewer. If the objective of the interview is to prevent the child from developing unhealthy drinking habits by exploring his drinking history and factors that contributed to the drinking, then the interviewer wants the parents to provide complete honest answers.

In the example just described, the question that minimized ego threats began with the contextual statement, "I know how hard it is to keep track of an active adolescent." In this phrasing, the contextual statement is used solely to demonstrate an understanding of the difficulties of raising a teenager and, thereby, to prevent defensiveness. Demonstrating understanding is an important means of maximizing empathy.

Empathy   Empathy is one of the critical components of effective interviewing, and examples appear throughout this book. Empathy is compassionate understanding. Empathy is *not* sympathy.

Empathic interviewers will strive to understand and communicate their understanding without becoming trapped in sympathy. Demonstrated empathy communicates, "I think I understand your feelings and hear what motivated your actions." Empathy does not necessarily reflect acceptance or agreement with the actions taken by the interviewee.

An empathic juvenile court counselor, who is listening to a preadolescent talk about the grinding monotony of school, can reflect understanding without agreeing to the preadolescent's solution of playing "hooky."

> "From what you are telling me, I can understand that you believe your teachers are boring, and nothing you learn in school is going to help you. Let's talk more about what you specifically find boring and see if we can figure out some ways to capture your interest."

Shorter empathic responses are helpful also: "So you feel ignored by your teachers?" "That sounds interesting! Tell me more about your role in that experiment."

The empathic contextual statement at the beginning of a question and the empathic response to an answer given by the interviewee bracket the interview to maximize empathy. Given the potential volatility of the interview, the interviewer may begin the interview with an introductory set of empathic contextual statements:

> "I know this situation has been very trying for you. I can tell you are still trying to get over the shock. Take your time answering my questions. I believe if we get all the information out on the table, we can have a better understanding of what happened."

The interviewer continues to maximize empathy by displaying genuine interest in the information given and then by summarizing what the interviewee says as the interview progresses.

Much of the effort to minimize threats to the interviewee's ego and to maximize empathy requires improvisation by the interviewer during the interview. However, during the preparation for the interview, the interviewer needs to predict and plan for areas that are likely to cause the interviewee to become defensive.

EXERCISE 3.3

Maximizing Empathy and Encouraging Recollection

Activity: Given a hypothetical interviewee and interview objective, the students prepare introductory, contextual, and empathic statements.

Purpose: To practice initial empathic techniques.

1. Interviewee: Probationer
   Interview objective: To discuss his or her progress on finding employment
2. Interviewee: Witness to an armed robbery who lives in the neighborhood
   Interview objective: To record what he or she saw and heard
3. Interviewee: Victim of a business burglary
   Interview objective: To ascertain mode of entry and reliability of current security system
4. Interviewee: Juvenile caught shoplifting
   Interview objective: To assess risk of future offending

Formulating Questions That Enhance Memory Retrieval

The interviewee may be willing, but unable, to give accurate information. Eyewitness reports often are unreliable (Shearer, 2005). Given the excitement of the moment, these so-called eye witnesses often describe what they think they saw rather than what they saw.

Witnesses may be feeling a high level and mixture of emotions, such as fear or shock, when they observe the event. If they are caught by surprise, then they may cognitively only register a limited part of the incident. (Convenience store operators have found that many of their cashiers cannot recall incidents of a robbery. To assist them, the doors to the store now have height charts on the frame, so at least they can get a better idea of how tall the robber might have been. In addition, more and more, often in today's economic downturn, store owners are installing 24-hour, closed circuit tape systems to assist employees in recalling incidents of a traumatic event, such as a robbery.)

Even if the witnesses carefully observed the entire incident, they might be confused about what they saw, they might repress the unpleasant memories from the incident, or they might not be able to articulate what they saw in a meaningful way. The interviewer needs to carefully and systematically recover what information the interviewee can retrieve.

Powell and Amsbay (2006) developed the *selectivity hypothesis* as an explanation of how distortion, inaccuracy, and incompleteness sneak into an eyewitness's retelling of an incident. These researchers credit memory distortions on the tendency of people to choose from a variety of stimuli at the scene during an incident to narrow what they observe. Complex events are too difficult to allow a witness to focus on the entire incident. Their selection often is influenced by experiences, attitudes, and expectations. Then, their retention is affected by stimuli that are favorable to their self-image. As memory fades, the interviewee might reconstruct the past by selective omissions, distortions, and fabrications.

R.P. Fisher has written articles and books (1992; 1995) on cognitive interview techniques to promote memory, especially for victims and witnesses. Fisher studied how police interviewed witnesses and concluded that a number of improvements could be made to facilitate the accuracy and completeness of the information received by the investigators.

Cognitive interview techniques have a number of presuppositions (Fisher, 1995):

1. All people have limited mental resources. The interviewer can minimize information-overload errors by refraining from interrupting with a battery of questions while the witness is searching through his or her memory for the information.
2. People remember more when they are in the same emotional state or place as when the event occurred, so recreating the original context is productive in retrieving memory.
3. The witness is instructed to mentally recreate the scene, including the emotions and physiological effects he or she was experiencing at the time of the event.
4. Retrieving the information from different approaches such as describing the event in reverse order or using different retrieval paths, such as visual, auditory, or tactile, can be helpful in recreating the original context.

Because the interviewee checks with the image presently in consciousness, it is more efficient to use that image to elicit all possible information before moving onto another image. So interviewees are encouraged to include every detail in their responses.

Fisher's techniques are based on two principles that enhance memory retention and retrieval: (1) the use of information from several sources to construct a recollection, and (2) the increase in accuracy when uninfluenced by external sources. Memory retrieval requires the interviewee to pull relevant information into conscious awareness. The interviewee then must convert the conscious recollection into a statement.

In helping the interviewee reconstruct the environmental, cognitive, physiological, and affective states that existed at the original time, the interviewer facilitates the complex retrieval process that the interviewee must go through.

Environmentally, the interviewee will be asked to detail the setting, such as the weather, furnishings, other people, smells, and noises—the elements necessary to recreate the original physical state.

Even before going through the actual content of the incident, the interviewer should develop questions facilitating recall of the interviewee's emotional and mental states at the time of the incident. The interviewer needs to be careful not to assign any emotions to the interviewee that the interviewer believes the interviewee should have felt. For example, the interviewer might think that a victim of a sexual assault is devastated, but the victim might have other life experiences that place a sexual assault at a different emotional level. The victim needs to describe his or her emotions, not have them dictated.

During training of sexual-assault-treatment personnel, one of the authors, Lord, recounted the following version of an incident from one of the students in the training:

> I work with child sexual abuse victims. Because of my training and experience, my local police department detectives ask me to interview rape victims for them, especially if they think there is something peculiar in the case.
>
> I was interviewing a sexual assault victim who described walking down the street one evening. A car pulled up with just a man driving, no passengers. He rolled down the window on the passenger side and said, "Hi, why don't you get in, and I will take you wherever you are going."
>
> The victim got into the car and told him where she was going. Soon after that, the driver pulled down a secluded street and stopped the car. He grabbed her arm and told her to take her clothes off. He didn't have a weapon nor did he hit her.
>
> She took her clothes off, and he raped her. He then told her to put her clothes back on. He drove down the street, stopped the car, and told her to get out. She did.
>
> The victim recorded the tag number and had a detailed description of the man, but her affect (emotional level, emotional expression, facial expressions) remained neutral during the reciting. She appeared as if she was a robot.
>
> I asked her, "Did you know the man?" Her reply was no.
>
> Then I asked her, "When was the first time a man ever ordered you to take your clothes off." Her reply, "When I was four years old."
>
> After a complete interview, she revealed that she had been sexually abused from the age of 4 to 14.

Victims, especially from an early age, remain victims all of their lives. Therapy helps and would help this victim. The victim described in the story did not reflect emotions normally seen by police when they interview rape victims. The investigators did not want to discount her report and had the forethought to call in an expert.

The physiological state of the interviewees also is important to know. Do the interviewees wear glasses or a hearing aid? Were they wearing their aids at the time of the event? Are they taking medications that might affect their observations or the memory of those observations? It is critical to be aware of anything that might have influenced the observations and the recollection.

Cognitively, the interviewer takes the interviewee systematically and slowly through every detail of what occurred, usually beginning with events leading up to the event, the event itself, and the aftermath.

> "I want you to slow down the moment that you first saw your assailant. Get a picture of him in your head. Now as you see him, describe him as completely as you can."
>
> She closed her eyes and after a moment, she began, "Well, it's dark so his face is in the shadows. He is wearing a knit cap that is striped. I'm not sure of the color. His ears are sticking out on each side. There are two gold rings in his right ear lobe. His hair is completely covered by the cap, and as far as I can tell, he doesn't have any facial hair.
>
> Oh! Wait! When he opens his mouth, he has terrible teeth-missing and crooked . . ."

Inexperienced interviewers are tempted to interrupt or show impatience while the interviewee is searching his or her memory. Interviewers must appreciate that it is incredibly hard and tedious work to recreate a detailed account of any event. The interviewer should remember to recognize and praise the effort that the interviewee is making and encourage him or her to report everything without mentally editing the recollections.

Fisher's second principle, or technique, explains the importance of reducing influences from external sources. External sources may cause witnesses to be unable to differentiate between what they actually saw at the time of the incident and what they heard about the incident later.

Perhaps the incident was reported on the television news broadcast, or friends discussed the situation with the interviewee. Interviewers should advise the interviewee to describe only what he or she saw. It is also useful occasionally to ask the interviewee explicitly to think about whether the critical event was actually observed or suggested by another source. Hearsay must be recognized and eliminated by the interviewer.

When uninfluenced by external pressure, most recollections are accurate. So, witnesses should not be directed to provide more information after they have indicated that they do not remember anything else. When witnesses are pressured to remember more, it is likely to increase the chances of getting inaccurate information. Instead, these witnesses should be encouraged to say they do not know or do not remember if they are unsure. The witnesses must understand that it is okay if they do not remember all the details.

The same problem exists with repeating the same questions. Repetition of questions is likely to increase inaccurate information. The interviewer is giving the message that an unsure answer is better than being nonresponsive. Children especially are susceptible to this type of pressure.

Other external sources include physical distractions during the interview. It is critical that the interviewer conduct the interview in a location in which all outside distractions are limited. Preparing the physical setting of the interview will be discussed in Chapter 4.

Fisher's cognitive interview (Fisher & Geiselman, 1992) is divided into five sections:

1. Introduction. In addition to building rapport and finding out a little more about the interviewee's background and interests, the interviewer conveys general guidelines to maximize memory retrieval and communication. The interviewee is reminded that the interviewer is not there to edit any of his or her thoughts. The interviewee is encouraged to concentrate thoroughly in retrieving from memory.

2. Open-ended narration. By using open-ended questions to encourage free flow of dialogue, the interviewee is allowed to complete the story untainted. The interviewer can develop an understanding of what the individual may know and what areas require further exploration by listening closely to ascertain how the interviewee's knowledge is stored.

   To recreate the general context, the interviewer explores the interviewee's state of mind prior to the event and the general physical environment surrounding the event. The interviewer will want to determine the interviewee's representation: what visual views of the scene and/or suspect did she or he have?

   To help the interviewee return to the scene mentally, the interviewer may want to ask, "What was the best view you had of the perpetrator?" or "Try to put yourself back into the same situation as when the crime was committed. Think about where you were standing at the time, what you were thinking about, what you were feeling, and what the room looked like."

   The interviewer should give the interviewee time to recreate.

3. Probing. As the interviewee is describing the event, the interviewer is developing a probing strategy: how to best bring to the interviewee's conscious images the best view of relevant information and then probe these images until all the information is exhausted.

   Because people cannot concentrate on more than one information source at a time, the interviewer should minimize nonessential sources of distraction, especially when the memory task becomes difficult and requires full concentration. Other views will be probed later. Because the process is so draining, the best view should be probed first.

4. Review. The interviewer reviews the information with the interviewee to confirm its accuracy and completeness. The interviewee is encouraged to listen actively and to interrupt if corrections or clarifications become necessary. The interviewer speaks slowly with pauses at each segment.

5. Closing. It is important that the interviewee is left with a positive last impression. The interviewer should request the interviewee to call if he or she recalls additional information and provide various ways of being reached.

As noted earlier, cognitive interview techniques encourage the use of different retrieval strategies to facilitate memory retrieval. Once the interviewee has given the events in chronological order, the interviewer should request that he or she tell it backward beginning with the last events seen. In this way, peripheral events, such as when the other perpetrator left the room, often are recalled. Also, backward recount can be useful when talking to suspects because they are often caught off guard, and it is harder for them to remember their lies and stick to their alibis.

Most people primarily remember visual information, although as discussed in Chapter 4, interviewees' dominant sense can be ascertained. Encouraging the interviewee to consider the situation through other senses, one at a time, such as probe for touch if physical contact was made, should be considered after the interviewer works with sight or the interviewee's dominant sense.

The use of contextual statements to motivate interviewees was mentioned earlier. Employing contextual statements to provide time and space perspectives increases the probability of accurate responses. If possible, phrase these statements to help the interviewee place himself or herself at the time and place.

> "I would like you to think back to around noon this past Friday, which was the fourth of June. I believe you would have been leaving your office for lunch right at noon because you said earlier the other assistant takes lunch at 1 P.M. Picture yourself at that moment and tell me what happened. Feel free to go very slowly and think about everything you saw and heard."

The authors of cognitive interview techniques suggest other strategies to retrieve specific information. These strategies suggest associating a feature or item with related attributes that may stimulate memory. For example, if the interviewee cannot remember a name, he or she might be able to remember if it was short or long, or how many syllables it possessed, which syllable was stressed, the beginning letter, if it was a common or unusual name, or a characteristic of an ethnic or national group.

Types of Questions

In general, questions are divided into open and closed questions. Each has its advantages and disadvantages, as well as appropriate uses. In addition, nonthreatening questions allow the interviewer to ask potentially threatening questions in a way they are perceived to be less menacing. There are a number of techniques surrounding nonthreatening approaches.

Every student knows the five Ws and H—who, what, when, where, why, and how—but there is definitely so much more to developing proper questions.

Open-Ended Questions   Open-ended questions are broad and most useful during the exploratory phase. The most basic open-ended question is "What happened?" This question encourages interviewees to talk freely and to provide the entire story chrono-logically. If the interviewees' emotional levels are high, allowing them to talk freely provides the opportunity to vent and decrease their emotional level. Visualize a seesaw—emotions and rational thought are on each end. When emotions are high, rational thinking is low. So as the interviewees' emotions defuse, their ability to think and to remember increases.

Questions surrounding *how* also usually solicit a full answer. An example such as, "How are you earning the money to pay your bills?" cannot be an-swered with one or two words, or a yes or no.

Using open-ended questions, the interviewer is able to explore the inter-viewee's specific feelings, biases, and vocabulary level. Using open-ended questions allows the interviewer the opportunity to hear the untainted story. Later in the interview, if the interviewee becomes reluctant to cooperate, the interviewer already has a great deal of information.

Of course, a disadvantage is that with a great deal of relevant information comes extraneous information that the interviewer must sift through to en-sure relevant and complete responses. The interviewee can drift away from the main objectives of the interview easier with open-ended questions, so the interviewer will need to stay alert and redirect the interviewee as necessary.

Closed-Ended Questions

Closed-ended questions can be answered with a short phrase and a finite answer. Frequently and optimally, closed-ended questions in the form of probes follow broad, open-ended questions to obtain specific information.

On the other hand, interviewers sometimes begin interviews with closed-ended questions especially designed to encourage responses from reticent interviewees. This approach puts these interviewees more at ease.

When asking closed-ended questions, the interviewer must be careful that the questions are not leading. Leading questions are not neutral, but rather tend to direct the interviewee's response in a specific direction. Leading questions most often contain the answer the interviewer is expecting. The interviewer does not want to suggest right answers under any circumstances.

An example of a closed-ended question is, "Did the robber have any dis-tinguishing marks?" A similar, but leading, question would be, "Did the rob-ber have a mustache?"

EXERCISE 3.4

Constructing Open-Ended Questions

Activity: Change the leading questions so they ask for broad, general information.

Purpose: To practice developing questions that will elicit valid responses.

1. How short was the person?
2. Was the man who was holding the gun African American?

3. Was the get-away car a blue Toyota?
4. Did you see the two juveniles who broke into the candy machine?
5. What color was the jacket that the assailant was wearing?
6. What jewelry did you see Officer Brown take from the store that was broken into?
7. How large was the semi-automatic that the suspect used in the robbery?
8. Did your cellmate's girlfriend bring in the cocaine on her person?
9. Which of the three men who robbed you also sexually assaulted you?
10. Who else has possession of the key to the back door where the perpetrator entered?
11. What color glasses was the robber wearing?
12. Describe the robber's tattoos.

Sequencing Types of Questions

The interviewer should consider mixing types of questions. It is common to consider the "funnel" sequence in which questions move from open to closed questions. For example, the interviewer begins with an open question to the victim, asking the description of the perpetrator. To fill in the gaps, the interviewer slowly narrows the questions to include the description of the perpetrator's clothes, type of shoes, and, finally, color of shoes. For shy or reluctant interviewees, the interviewer may consider inverting the funnel, or asking closed-ended questions first and then moving to more open-ended questions.

Detectives often begin by asking neutral, closed-ended questions, such as the interviewee's complete name and contact information, especially to victims to facilitate their comfort level. Neutral, close-ended questions are helpful in de-escalating highly irate disputants. As the disputants begin to answer straight-forward questions, they release some of their strong emotions.

Each interview must be deliberated before selecting the sequence of questions. Beginning with closed-ended questions should be considered carefully and cautiously. Two or three neutral, closed-ended questions may allow a victim to become more comfortable or allow an angry complainant to calm down. Too many close-ended questions can either increase the anger or frustration of the interviewee or begin a pattern of leading questions that do not allow complete or valid answers. Beginning with open-ended questions is usually encouraged to allow as much of the complete story as possible to be told without interruption.

The interviewee has information that the interviewer wants, so the majority of the talking should be the interviewee's responses. The interviewer should keep this statement in mind at all times. A large number of closed-ended questions may lead to the interviewer talking the majority of the time during the interview instead of the interviewee.

Nonthreatening Questions

No matter if the interviews are investigative or helping, interviewees will not be motivated to respond honestly or completely if they feel threatened. As discussed earlier, the threat may be to their self-esteem or freedom. Interviewers must learn to ask questions in such a way that they promote flowing narratives that reveal all the elements of an incident and the interviewees' involvement.

In the case of helping interviews, nonthreatening questions will be framed in nonjudgmental, problem-solving language. The interviewee can feel the compassion and appreciate the interviewer's avoidance of hasty conclusions.

Charles Yeschke (1997) has developed a variety of nonthreatening questions to be used for potentially reluctant witnesses and suspects. Throughout the descriptions of these questions, the establishment of curiosity and desire to understand is prevalent.

Indirect Questions

There are a number of questions that provide useful information without directly pointing a finger at the interviewee. One type of indirect question asks the interviewee to discuss the type of person he or she believes could have committed the act. This general question often provides, subconsciously, actual characteristics of the responsible individuals if the interviewee either knows the offenders or is directly involved.

Slightly more direct questions then follow, relating the characteristics with suspicions toward other individuals the interviewee knows. The interviewer carefully phrases the questions so the interviewee understands that he or she is not "ratting" on another. At the same time, the interviewer receives additional information from the response.

For example, the interviewer introduces the question, "You said that you don't know who took the money from the cash register, and I don't want you to feel pressured to accuse anybody. What characteristics do you think the person would need to have to take the money?"

After receiving a response of the type of characteristics, the interviewer would ask, "Is there anybody you work with that fits those characteristics?" The interviewer might add, "Think back over the past three or four days. Was anything said or did anybody do anything that might fit those characteristics?"

A follow-up, slightly more threatening question is, "Is there any reason that another employee would believe you were involved with the disappearance of the money?" This second, more direct question still does not target the interviewee. At a minimum, whatever the response, the interviewer has elicited additional information that may be useful. An interviewee who is not involved in the theft usually thinks a moment and suggests either other witnesses for the interviewer to question or a process that had not been mentioned.

For example, the interviewee might state, "I don't think so. However, I am responsible for bringing change to the cashiers. I always follow the policy that states that I count the change directly into the cashiers' hands, and they place it in their cash registers. You could check with all of the cashiers and see if I ever deviate from that process."

Although the interviewer probably has been given the names of all of the cashiers, the list of cashiers could be checked by requesting the interviewee to provide the names of the cashiers working during the time the money disappeared.

On the other hand, an interviewee who is involved with the theft will feel the need to project the suspicion toward other targets and, therefore, makes a general response. While the response may be general, it will still provide additional routes to ask further questions.

> The interviewer asks, "Is there any reason that another employee would believe you were involved with the money's disappearance?"
> The interviewee responds, "All the other employees around here would think I was involved."
> Interviewer: "What would make them think that you were involved?"
> Interviewee: "They don't have any reasons, but they don't like me because I don't hang around with them."

Another type of indirect question originates from the opposite direction—that of trust. The interviewer asks, "Who do you know who couldn't have possibly taken the money?"

Although some interviewees may provide a flippant response such as "my mother," most interviewees feel unthreatened by the question and give several names.

The next question then becomes, "What is it about those folks that would make it impossible for them to be involved with the disappearance of the money?"

Hypothetical Questions

There are a number of ways to ask what if questions, allowing individuals to discuss what they know, but because the question is framed as a hypothetical, it is perceived as nonthreatening.

One of the most famous series of interviews using this method was employed by Stephen Michaud and Hugh Aynesworth (1984), two *Miami Herald* reporters, who interviewed serial killer Ted Bundy. The reporters realized Bundy would never confess, but they could tell he wanted to brag about his murders. So the reporters used the hypothetical question, "How do you think the killer committed the murders and escaped apprehension?" Although Bundy never confessed, his long interview described how he would have conducted the murders if he were the killer.

> Ted . . . began describing the killer. "Within this individual," Bundy explained, "there dwelt a being"—Ted sometimes called it "an entity," "the disordered self," or "the malignant being." The story of its beginnings came slowly, chronologically, a consistent tale of gathering sociopathy that nurtured itself on the negative energy around it. (Michaud & Aynesworth, 1984, p. 20).

It is commonly known among professionals that thinking about an act does not make one guilty of the act, so to ask, "I know it must get tedious sometimes working third shift with time to fantasize about doing things. Have you ever found yourself thinking about what you could do with a couple thousand extra dollars—of course if you didn't get caught?" (laugh)

Or perhaps, the question, "Have some of the other guys around here ever approached you about diverting a few dollars?" is not particularly threatening. Individuals regularly think about winning the lottery, but possibly strong reactions to the questions might provide areas to explore further. A follow-up question could be, "How do you think it could be done?"

A general hypothetical question such as, "You know this business upside and down, inside and out. If you were one of the individuals who took the money, how do you think you would do it?" can often reap bountiful information without pointing a direct finger at the interviewee. Remaining with that type of question, the interviewer could ask, "Why wouldn't they get caught?" or "How would that work?"

Hypothetical questions also work well in helping interviews in which the interviewer–helper is assessing the interviewee's ability to change behavior or cope with problems, for instance, "What if you were to leave the halfway house like you are considering? Where do you think you would find a place to live?"

Conditional Questions

Beginning questions with *if* softens the impact of the question, especially if the interviewer's manner is open and empathetic, showing real interest rather than judgment.

Conditional questions can range from, "If you are involved with this incident, I know you want to clear up your role as quickly as possible" to "If you are having problems staying away from associates who use drugs, let's talk about it now so we can figure out a means to prevent any actions on your part that could send you to jail."

Control Questions

Zulawski and Wicklander (1993) suggest the use of control questions to observe differences in the behavior of the interviewee. For example, if the interviewee is a potential suspect, a direct question about the offense will be asked and his or her reactions observed while denying any involvement. Then, the interviewee would be asked a question such as, "Have you ever violated agency policies that you would prefer your supervisor did not know about?" If the interviewee shows more concern and more change in behavior with the violation of a policy question than the question about the offense, then it is probable that the interviewee is not involved with the offense.

"As I know you have been told, we are asking all employees about their knowledge about the theft of the money from the office that occurred yesterday. Did you take the missing money or have any involvement in the money's disappearance?"

"No, I did not." (calmly stated)

"Have you taken any money from the business?"

"No, I have not." (calmly stated)

"Have you violated any of the business' policies that you would not want your supervisor to know about?"

"Well (hesitation, looking away, chewing on a fingernail), I don't think I have ever done anything that was really wrong, but my supervisor might consider it a policy violation."

Even though the interviewee did not want to directly discuss what the policy violation was, the change in nonverbal behavior and hesitation demonstrates his or her concern for getting reprimanded for an actual misbehavior. This noticeable change of behavior indicates that the interviewee is not capable of controlling behavior that would make him or her a good liar and is likely to be telling the truth about the actual offense.

Questions to Avoid

Why Questions   Although included in the five Ws and H questions, why questions should normally be avoided. It often comes across as judgmental and sets the interviewer up to be perceived as condemning the interviewee. Not only does asking why imply blame and disapproval, it is likely to cause the interviewee to rationalize or defend his or her actions, to withdraw, or to attack because of feeling threatened.

Although the techniques in the Reid Nine Steps of Interrogation (Inbau, Reid, Buckely, & Jayne, 2004) include the interviewer developing motivation for the crime to move a guilty person to confession, it is a questionable tactic (Gudjonsson, 1993). The developed motivation for the crime is actually providing rationale for the criminal behavior. Developing motivation during interrogations will be discussed in Chapter 10.

Behavior is usually driven by complex motivations that are not really understood by most interviewees, even those not charged with a crime. When asked *why* they committed a particular act, most individuals will look bewildered and state they do not know. The accompanying discomfort may turn them off from answering any additional questions.

In a few cases, why can be used for simple facts such as, "Why did you decide to take the courses in plumbing?" Even with these simple, fairly innocuous questions, a trusting and respectful relationship between the interviewer and interviewee needs to have been established before the interviewer strays down that path.

Questions That Lead to Confusion

Rapid-fire and double-barrel questions create incorrect answers, confusion, and an interviewee who is likely to withdraw.

Questions should be asked slowly and thoughtfully with interviewees given time to complete their responses. Rapid-fire questions look good on television, but are taboo in interviewing. The interviewee does not have time to fully explain or even catch a breath.

Double-barrel questions shoot out two or more questions in the format of a single question. One of the most famous comedians asks this double-barrel question: "Have you stopped beating your wife?" Either way, the interviewee tries to respond and provides a negative answer. A negative response is interpreted that the interviewee has not stopped beating his wife rather than being interpreted as denying a behavior never used by the interviewee. An affirmative answer implies that he has stopped beating her after having beaten her for a period of time.

Leading Questions

Leading questions in which the interviewee is directed toward a specific answer is often unintentionally asked when developing closed-ended questions. Intentionally asking leading questions also belies ethical questioning and is a practice that interviewers need to guard against.

The ethical goal of interviewing is to obtain valid information. Often interviewees, especially children, are trying to figure out what answer the interviewer wants. Any question that has the least bit of tilt to a particular answer may induce the interviewee to give that specific answer rather than the valid answer. Even questions as harmless sounding as, "Have you remained sober this week?" is likely to lead to the answer the interviewee believes the interviewer wants to hear—yes. If one of the conditions of probation is to remain sober, a better question would be, "Let's discuss your probation conditions and see how you are progressing. How did you do this week towards reducing your use of alcohol?"

Loaded questions in which emotional words are used such as, "Why are you lying?" or "When did you first start abusing your children?" can cause the interviewee to react defensively and shut off communication. Loaded words are judging words that appear to be directed at the individual. These words include *crazy*, *thief*, and *rapist*.

Likewise, asking sadistic "cat and mouse" questions such as, "Do you know why you were asked to come in and talk?" does not facilitate communication, but rather generates hostility. The interviewee knows that the interviewer's intentions for asking the question are not to gather information, but rather to increase the interviewee's anxiety and/or other emotions.

With the numerous possibilities for constructing a variety of effective questions, interviewers should easily be able to avoid unhelpful questions.

Conclusion

This chapter outlines the effort needed to achieve an effective interview. The interviewer must review available information and learn as much as possible about the interviewee. A well-defined objective or objectives must be specified so that relevant questions can then be constructed. Returning to the objective at the end of the interview also helps the interviewer assess the effectiveness of the interview.

The interviewer needs to think about the types of questions that are likely to obtain valid, complete, and relevant answers. Means to enhance memory and reduce perceptions of threat to the interviewee's self-worth are an important part of question development. Formulating questions that will encourage responses and enhance the interviewee's memory are essential skills.

Patience and empathy are important virtues to cultivate. Although planning interviews appears to be complex and tiresome, once background information is compiled, experienced interviewers are able to develop objectives and relevant questions fairly quickly. Just as with any skill, interviewing preparation takes practice.

EXERCISE 3.5

Constructing Questions for Objectives of Students' Interviews

Activity: Carefully writing questions to which responses should meet the objectives of the students' interviews.

Purpose: To practice writing relevant, nonthreatening questions and to receive feedback from other students and the instructor.

At the completion of this chapter, students write questions that they believe are relevant for the objectives of the interviews with criminal justice professionals. The questions should be organized and primarily open-ended to facilitate complete responses. The students should consider sensitive areas that might be considered threatening to the interviewee and construct the questions in a way that they believe will result in valid answers.

The instructor may want to consider the students exchanging their questions with each other for peer feedback and discussion before providing his or her own feedback. The students' interview objectives should be included with their questions.

ALLEN'S WORLD

Dr. Lord's world and my world overlap in many areas. Preparation is the key not only in my 20 years as a private investigator, but a previous 20 years as an investigative reporter.

In both fields, I rarely use staccato questioning. I clearly have an objective, and it is well defined. It cannot be stressed enough that, regardless of the type of interview, there are only three things that matter: preparation, preparation, and preparation.

Let us begin with investigative reporting.

Frank Kisler, a black activist in the 1970s wanted to develop a planned community called Soul City for African Americans in Warren County, a rural North Carolina tract of 5,000 acres near Manson. Kisler got $14 million from the federal government funding source, the U.S. Department of Housing and Urban Development (HUD), and a lot of private donations.

Sad to say, the project was not going well, and outsiders believed there were misappropriations afoot.

I was assigned to interview Kisler about Soul City in the mid-70s.

I began in the "morgue," a journalistic term for the library. Newspapers used to clip all articles about a person and subject, file them, and use them for reference.

I got Kisler's file. Read and memorized much of the information. How much money did he get from the feds? What did his plat map look like? Whose names were on the deeds? Who were the subcontractors, how much were they paid, and were the contracts put out for bid? What had been his activities with The Congress of Racial Equality (CORE).

I called Kisler, made an appointment, and drove up to see him.

Some interviews are routine. Other interviews, the interviewee has a good idea of the questions and has prepared rote responses. Kisler was one of those who prepared rote answers.

To get an effective interview, I had to shock him. Jolt his "tape" so to speak.

"Why hasn't somebody killed you?" was my first question. Kisler's jaw dropped. His eyes told me his tape had just gone berserk. He was thrown off balance, which produced a much better interview.

I then began with open-ended questions. How is the project going? Are you pleased? Basic "softballs over the heart of the plate" questions. Hit them out of the park for home runs, Mr. Kisler.

Then, the important questions . . .

Where has all the money gone?

How many family members are on your payroll? What do they do? Are they here now? Can I talk to them?

Where's the infrastructure in Soul City? Why are there not any houses? And on, and on, and on.

At the end, it was clear Kisler had spent a lot of money with little results. The key was my preparation for the interview.

In June 1980, the federal government assumed responsibility for Soul City.

Now, let us take a look at how I prepared to interview the president and CEO of the PTL (Praise the Lord) evangelistic television network.

In late 1978, I learned from sources that PTL's CEO was raising money on television to build missionary programs in Brazil, Cyprus, and South Korea, but he was spending that money at Heritage USA in South Carolina, just over the border from Charlotte, North Carolina.

In this type of interview, the objective is clear: prove and document the misappropriations. I needed help, so I enlisted another reporter to work with me.

We went to see Robert Blessing, a former vice president of PTL who had resigned. We had a friendly interview, a lot of open questions, a lot of closed-ended questions, and one critical leading question.

Dr. Lord says that you should avoid why questions. But one of my first questions to Blessing was, "Why did you resign?" Blessing was reluctant to pinpoint his precise reason for resigning.

In this case, I felt it was okay to ask a why question because Blessing had already resigned and mentally had decided it was in his best interest to leave PTL. The why question was designed to get him to elaborate his reasons for leaving, which would ultimately get me to my main goal—finding out how the CEO was running PTL.

"Robert, can you give me some examples of things you are unhappy about?" That is another example of an open-ended question.

I used close-ended questions to clarify his answers to some of my open-ended questions.

For example, Blessing said he was unhappy that the president, at times, would not listen to his advice. At that point, I asked a closed ended question: "How long has the president been ignoring you?"

And then I asked the critical leading question. Robert told me that the president was going on television asking for money for missionary construction. Robert knew the money was not being used for missionary construction.

So I asked a leading question: "Is the president taking money from the ministry that doesn't belong to him?" I felt justified in asking that leading question because I was not putting ideas in Robert's head, which is the danger of leading questions. I believe in my investigative work today that it is okay to ask a leading question when I am convinced that I am not supplying information or putting ideas in the individual's mind.

Next up, I went to interview William Parker, PTL's former financial officer who had resigned. It was a friendly interview with a lot of open questions, and a lot of closed-ended questions. And then, the critical question: "William, can you document the misappropriations?"

Parker had ledger sheets showing exactly how much money had come in earmarked for the three specific overseas projects. Parker gave me the ledger sheets.

Next up, Carl Masters, a former vice president who had resigned. It was a friendly interview with a lot of open questions, and a lot of closed-ended questions.

All three former executives— on the record and for attribution— confirmed that money had been misappropriated from PTL by the President. The next step was to call the ministers in each of the three countries that PTL's president had promised to help. This meant we needed translators available. We called South Korea and Cyprus about midnight

Charlotte time. All three ministers confirmed they had appeared on the PTL television network and made appeals for money to build their ministries. None of the three ministers ever got a cent from PTL, not, at least, until our disclosures were printed.

At some point, PTL did send out token amounts to show it was trying to appropriately distribute the money that had been raised through the televised appeal. It was too little, too late.

The next step was to get tapes of those broadcasts.

The process took months. Finally, we went to interview the CEO.

Here is where we violated Dr. Lord's guidelines. In this case, we knew most of the answers to every question we were going to ask PTL's CEO. They were not open or closed-ended. They were directed questions—threatening questions. We started off with the key question: Not, "Did you misappropriate money sent in for overseas ministries?" But, "Why did you misappropriate money sent in for overseas ministries?"

The CEO was discombobulated and he said: "How did you know?" which for us was confirmation.

In January 1979, the first article ever published about PTL's financial misconduct appeared on the front page of *The Charlotte Observer.* Almost immediately, the Federal Communications Commission (FCC) began an investigation into PTL's financial labyrinth—preparation, preparation, preparation.

Preparing interviews as a private investigator often is different than being a reporter, but still, a lot of the preparation and interviewing conflicts with the nonthreatening interviews described in Dr. Lord's academic world.

Preparation is still the key. The style and tenor of interviews cross a broad spectrum. What that means is, when I arrange the interview, I know the answer to almost every questions I am going to ask, before I ask it. It is the same mantra for attorneys. Never ask a question of a witness on the stand unless you know the answer in advance.

Let us talk about leading questions. Dr. Lord is absolutely right—always avoid, at all costs, asking a leading question, except for the exceptions. Leading questions are improper because they always are directing the interviewee to give a response desired by the questioner. If the questioner knows that the interviewee knows the answer—and is confident he or she is not supplying information to the questioner—I find it acceptable to ask a leading question. Again, if my preparation is thorough, I already know that the person knows the answer. I am just prompting him or her to remember with the leading question. I do not ask leading questions often. But from time to time . . .

Let us look at a few of my cases, and how they verify Dr. Lord's guidelines.

I have a client who is in jail on a first-degree murder charge. The main witness to the shooting drew a picture for the police of the parking lot where the shooting occurred and where he was standing when he saw the shooting. I went to the parking lot. The police had marked the spot where the witness was standing. I noticed on the asphalt there were three heavy black lines on the ground, like a rectangle, only one side of the rectangle was missing. Those markings told me a dumpster used to be in that spot. I found the apartment manager and asked a leading question: "When was the dumpster moved?"

The dumpster was moved weeks *after* the murder, which meant the witness was lying. He could not have seen the shooting. I found the witness and asked the opening question: "Why are you lying about what you saw?" It turns out the key witness was actually one of the shooters trying to deflect blame. When his story collapsed, the district attorney dismissed charges against my client.

In another case, my client is charged with sodomizing a male friend who he got drunk one night. The alleged victim gave a detailed statement to the police. Preparation—do a criminal background check. Has the alleged victim filed any civil complaints claiming he has been abused previously? Who had he worked for and what was his reputation?

In his statement, the alleged victim said he was given a date-rape drug before the assault and went to the Rape Crisis Center for help after the assault. I wanted to know if he went to the hospital for an exam.

Here is what I learned. The victim went to the hospital two days after the alleged assault so reliable testing was impossible. The Rape Crisis Center does not have walk-in traffic. When victims call the center, they are told to go to the hospital, and a Rape Crisis counselor meets the victim. The Rape Crisis folks had no record of the victim calling for help.

The victim had filed a workmen's comp claim against a former employer. The former employer said the claim was "dubious."

I never got to interview the victim.

During the process, the victim asked for a huge settlement to have the charges dropped. Eventually, my client paid rather than take a chance with a jury trial.

EXERCISE 3.6

Putting it Together

Activity: Beginning first with writing the objective of the interview, the students then draft a list of questions to ask the alleged sodomy victim in Allen's World.

Purpose: To integrate interview preparation steps

After reading Allen's World, the students individually or in small groups construct the objective(s) of the interview of the alleged sodomy victim/complainant, and then develop questions that are relevant to the objective(s). The individual students or small groups report their objectives and discuss the relevance of their questions and potential responses.

PART

Conducting the Interview

Beginning the Interview

Upon completion of this chapter, the student should be able to

1. Describe the phases of the interview
2. Establish rapport
3. Actively listen to responses
4. Use nonverbal communication as an additional interviewing tool
5. Prepare a suitable physical setting for the interview
6. Understand what is needed for a suitable verbal setting for the interview
7. Prepare and complete the beginning elements of an interview

The detective looked over the office. The two chairs were situated about six feet apart, facing each other, but each slanted about five degrees toward each other. The detective had put out the group picture of a mission trip he had taken last summer and made sure that the United States flag was seen in the corner. The blinds had been pulled and the chairs were back-lighted. There were tissues on the side table nearest the interviewee's chair.

A knock at the office door.

The detective walked over and opened it. A father and teenage daughter stood uncertainly. The detective held out his hand to the father, "Mr. Harris, thank you for coming down to talk to me. As I mentioned on the phone, I'm Detective Burton with the Family Services Division of the Metro Police Department. I've been working cases involving children for ten years. From what you said on the phone, I think I can gather some information from your daughter, Margaret, right? [The detective smiles at the teen, but decides not to attempt a handshake] And be able to get things straightened out."

The detective extends his hand toward one of the chairs, "Margaret, please sit here. Mr. Harris, if you don't mind, I would like to ask Margaret a few questions. Why don't you have a seat in the main office?" Mr. Harris looks from the detective to Margaret. Margaret shrugs and sits down. Mr. Harris walks out. The detective leaves the door slightly open.

"Margaret, I have told the secretary to make sure we are not bothered. Can I get you something to drink? Do you need anything before we begin?" Margaret shakes her head no.

"Margaret, hopefully you won't be uncomfortable with the personal questions that I need to ask. I will explain the reason behind my questions and give you plenty of time to gather your thoughts. I will be taking notes, but have decided not to tape record our conversation at this time. While I can't promise you full confidentiality, my notes are only available to my supervisor and the district attorney assigned to the case. They may be subpoenaed at some point down the road."

"Margaret, do you have any questions for me at this time?"

Introduction

Although interviews vary in their purposes and length, they always have three phases or parts. The purpose of each phase is dependent on whether the interview is an investigative or a helping interview. As discussed in Chapter 3, the interviewer needs to formulate the objectives for each interview. Then, it is possible to anticipate what can be achieved at each phase of the interview.

From the first introduction, relationship building must be developed between the interviewer and the interviewee. The interviewee needs to feel comfortable and to develop at least a minimal level of trust with the interviewer. The interviewer establishes a relationship with the interviewee through gaining rapport and actively listening. There are a number of techniques linked to these two vital skills.

There has been a great deal of publicity about nonverbal communication, and it will be discussed in this chapter. First to be considered is the interviewer's nonverbal communication and what it conveys to the interviewee. The interviewee's nonverbal communication will also be discussed in depth. Because of the amount of hype that has been publicized about nonverbal communications, especially around deception, it is important the reader learn about the factors that should be considered before attempting to interpret nonverbal communications.

Besides preparing questions before the interview is conducted, the interviewer must take time to prepare the interview setting. To the extent possible, the interviewer controls the physical setting; however, the verbal setting also must be considered. Engaging interviewees immediately by addressing some of the unspoken questions and needs they have will contribute to an effective interview.

Interview Phases

Although the phases of investigative and helping interviews differ, all interviews should be planned based on what the interviewer needs to accomplish at each phase. Such planning will help the interviewer organize the interview and ensure the completion of the objectives.

Phases of the Investigative Interview   Charles Yeschke (1997) divides an investigative interview into three sections: initial phase, primary phase, and terminal phase. The initial phase is further divided into precontact, strategic planning, and initial contact.

Initial Phase

The objectives of the initial phase are to prepare for the interview. As emphasized in earlier chapters, ideally the interview should not take place until the interviewer has compiled all the known information about the interviewee and associates and the precipitating incident.

During precontact, the interviewer becomes familiar with the available information about the incident and/or the specific interviewee. What information has been gathered from other interviewees? What evidence has been collected? Was information obtained from a canvas of the neighborhood? Does the interviewee have any disabilities, perhaps a hearing impairment? As the information is gathered, the interviewer can simultaneously begin to formulate a strategic interviewing plan. For example, what will be the best approach to motivate the interviewee? The interviewer can begin to develop questions, deciding which questions will help confirm information already gathered and what information will explore possibly unknown territory.

Even with preliminary interviews of victims, the interviewer can review whatever is known before his or her interview. Then, before returning for a follow-up, in-depth interview, the interviewer should have completed the initial phase that includes reviewing the physical evidence from the crime scene, information from the neighborhood canvas, or other related case information.

It is important to understand the relationships among any involved individuals and, if crime related, which individuals had sufficient knowledge, opportunity, access, and motive to commit the crime. In regards to the key individuals who will be interviewed, background checks should be conducted that include information connecting them to the possible crime, as well as information about their associates, personalities, hobbies, and habits. The more facts that are gathered on each interviewee, the more likely that the interviewer can build rapport and avoid topics that might create defensiveness.

The interviewer should approach every interview with the positive expectations that the interviewee will provide valuable information; however, a realistic assessment of the probability of obtaining a valid, complete interview is useful in planning the interview strategy. The interviewer needs to be prepared to include the necessary motivation to obtain a useful interview. During the strategic planning stage, the interviewer will clarify the objectives for the interview and construct questions. Although the experienced interviewer will be able to improvise prompts that are needed to further clarify and expand the interviewee's responses, the construction of questions during the planning stage will ensure a complete and relevant interview.

The last subphase of the initial phase is the first 4 minutes of the actual interview in which the primary purpose is to establish rapport. The first impressions are critical. From the first moment, the interviewee starts to evaluate the interviewer and to decide whether it is safe to speak. The physical setting is an important component that the interviewer should plan. It will be discussed in more detail later in the chapter. The information that has been gathered to facilitate rapport building will help set the tone of the interview.

As the interviewer states the objectives of the interview and expresses appreciation for the interviewee's cooperation, the interviewer will be receiving his or her first signals from the interviewee of the need to vent emotions of fear or anger, a desire to be helpful, or the need to be deceptive. First impressions really are powerful, but the interviewer also needs to be careful to integrate the first impressions with each new piece of information obtained.

Returning to the opening scenario with the father and daughter, the interviewer briefly provides information that helps establish credibility and also polite authority to allow an interview with the daughter alone. The interviewer then focuses on making the girl as comfortable as possible given the circumstances.

EXERCISE 4.1

Introduction

Activity: Videotaping students as they role play the introduction they plan to use to conduct their interview.

Purpose: To help the students become more aware of how they present themselves, to emphasize the importance and complexity of the introduction of the interview, and to allow students to practice observing other people's nonverbal cues.

The instructor should set up the interview room with two chairs, a table, an audio recorder, and tissues. Each student should be allowed time to set up the room and express how he or she wants to begin, e.g., the student interviewer already is sitting in the room, and the interviewee enters.

[Note: The instructor may want to emphasize the frequency and importance of field interviews and include settings outside with students standing.]

The session is videotaped for about 5 minutes. During that period, the student interviewer should introduce himself or herself, include one or more rapport-building statements, and state the objectives for the interview.

After the interview, the instructor has two teaching options:

1. Replay the video segment. Ask the student interviewee to discuss how he or she felt toward the interviewer. Was it clear what the interview would be about? How comfortable did he or she feel with the interviewer?

   Ask the interviewer to discuss his or her strengths and weaknesses. What will he or she do differently when conducting the actual interview or next interview?

   Ask the students who were observing to give additional input on what they saw as strengths and weaknesses.

2. After acting out the introduction, the student interviewer answers briefly in writing the following questions:
   a. How do you think the interviewee felt at the beginning of the interview?
   b. How do you think the interviewee felt by the end of the interview?
   c. Describe the interviewee's nonverbal cues. What were his or her facial expressions? What did he or she do with his or her arms and hands, legs and feet, and posture?
   d. What do you think are your interviewing strengths?
   e. What do you plan to practice and improve?
   The student interviewee answers briefly in writing the following questions:
   a. How did you feel at the beginning of the interview?
   b. How did you feel at the end of the interview?
   c. What do you think the interviewer did well?
   d. What do you think the interviewer should do differently?
   The student interviewer and interviewee replay the video segment after they share their answers with each other and the class.

Primary Phase

Yeschke (1997) considers the primary phase the period in which the major part of the information gathering occurs. Chapter 5 details how the interview should be conducted. For the purpose of preparation, the interviewer develops a set of questions that he or she believes will elicit the needed information. Chapter 3 describes the development of interview questions.

The questions also will tease out any inconsistencies within and between interviews. Once the interviewer believes that all of the information that is possible for this interview has been obtained, it is time to ask the interviewee to help summarize the information.

Terminal Phase

The purpose of the terminal phase is to assess the need for follow-up interviews and to summarize the information. If necessary, after the interviewer compliments the interviewee on his or her hard work, the interviewer suggests a brief break so the interviewer can evaluate if the objectives of the interview have been met. Has the interviewee provided complete, relevant, and valid answers to all of the interviewer's questions? Does the interview need to continue? If there are discrepancies that need clarifying, what are the probabilities of the interviewee allowing another interview? Is it more important to continue and risk the interviewee becoming increasingly reluctant and tired?

An important component of the terminal phase is the need to maintain rapport throughout. Even if the interviewer believes that all of the possible information has been obtained from the interviewee, in all likelihood, a follow-up interview will be needed. If the interviewee feels respected and recognized for his or her contributions, it is highly likely that the interviewer may receive a positive response to "If you remember anything else, please contact me. Don't worry or try to decide if the information is important or not; I will make that decision. You have been so helpful."

While summarizing the information, the interviewer should break it up into manageable components. To keep the interviewee involved, the interviewer should ask for confirmation of the summary's correctness and add small clarification questions and comments regarding the importance of the information.

> "Margaret, I know this was difficult for you. It was very courageous for you to come forward and tell me what happened. I want to go over what you said so that I am sure I got it correct. Please correct me if I got any of it wrong, ok?"
>
> "Now, I believe you said that you first met Coach Graham during try-outs this past spring after spring break. Was that around the middle of April? . . ."

Phases of the Helping Interview

Krishna Samantrai (1996) divides helping interviews into beginning, middle, and ending phases. The beginning phase assesses the interviewee's needs and begins the relationship building between the client, or interviewee, and the interviewer, or helper. The middle phase is considered the intervention component. The ending phase terminates that specific interview and establishes the link for the next interview when needed or required.

Samantrai (1996) notes that the three phases are not discrete, separate processes, but continuous and overlapping. Relationship building must continue throughout all of the phases. Assessment and intervention are circular and iterative. As intervention occurs, the client is reassessed.

Before meeting with the client, similarly as with the investigative interview, the helper should have a clear idea of the purpose for the interview and who the client is. During the initial phase, the helper introduces himself or herself and states the purpose for the interview. During this introduction, the client should be encouraged to provide feedback about the interview process and the purpose for this and additional sessions, to ask any questions, and to respond to questions from the helper. Confidentiality and its limitations according to law and agency policy will also be explained during the beginning phase.

Once rapport is established, and the client appears comfortable, then the helper will begin to assess the client. The client begins by describing the presenting problem, which the helper explores and clarifies with the client. Assessing the extent and duration of the problem and its impact on the client and how the client has attempted to remedy the problem is critical, initial information.

The helper then explores the client's history, family, social environment, daily functioning capability, internal strengths, resistance, external support systems, and possible barriers to intervention. Clients exist within the context of their social environments; the relationships among individuals, families, and society are reciprocal and dynamic. The helper needs to get a sense of the client's level of functioning and how well he or she meets the demands and expectations of various roles. Quite often agencies have special assessment instruments that are administered to the clients by trained personnel.

During the middle phase, the client begins solving problems with the assistance of the helper. Interviewing becomes secondary to problem solving. The middle, or intervention, phase is an action phase. Many clients' needs can be overwhelming and may need to be divided. In other words, with the helper's assistance, the client makes a list of each category of need, lists the tasks involved in each category, and then decides how each task will be done within a time frame. Broken into small parts, the tasks feel doable and keep the client from feeling overwhelmed. The helper's role is to support and help structure the problem-solving process that the client must go through as independently as possible. It is useful for the helper to be knowledgeable of community resources.

The ending phase terminates each interview session. The client should be clear on how long the sessions will continue. Unless there are only one or two sessions, the helper should set up a ritual for closure. Formal leave-taking should include summarizing what was discussed and accomplished during the session and clarifying, with the client's help, what will happen next.

> "George, I think you accomplished a great deal today. You came in feeling pretty blocked by a problem, and you are leaving with a plausible solution that you feel capable of implementing. What do you think we should work on next week?"

Structured phases are helpful for interviewers and the interviewees. During investigative interviews, knowledge of the purpose of each phase facilitates the organization of the interviews and ensures a complete interview. In helping interviews, the client is given structure that he or she can learn to depend on, and which also helps the helper construct a beneficial interview and session.

Establishing Rapport

> She nervously looked around the office. How could she talk to this stranger about being raped?
>
> The detective had a picture of a sailboat on one wall, a pair of dirty running shoes draped over a chair, and several different family pictures with children of different ages. (Unbeknownst to her, the detective, after years of interviewing, had figured the type of props that would help him connect with people he might be interviewing.) She smiled slightly when she noticed one of the children looked about the age of her little sister.
>
> "Is that your daughter?" she asked.

"Yes, she has just started the fourth grade. Boy, does she have her own mind!" the detective replied.

"She sounds like my little sister. My sister is starting fifth grade."

. . . And the rapport building begins . . .

Rapport building equals relationship building; two people connect. The interviewee needs to feel comfortable with and trust the person who is asking the questions. Rapport building does not require emotional involvement, but rather psychological closeness. The interviewer needs to project trustworthiness, professionalism, and knowledge gained from experience.

As noted above, finding a common interest or a shared concern draws the person out and reduces resistance. Once he or she begins talking about these common areas, a comfortable feeling and a decrease in any apprehensions occur. It is important to project understanding and an appreciation for the interviewee's feelings. This projection emerges first by sharing common areas and then expands as the interviewer shows he or she is listening. The initial duration of rapport building is brief. The interviewee should begin to relax and talk a bit about himself or herself almost immediately.

Research (Collins, Lincoln & Frank, 2002; Doerner & Lab, 1998 as cited by Homberg, 2004) has found that interviewees who are approached initially with rapport building are more likely to provide complete answers and stay with the investigation through the judicial process.

Victims with whom rapport has been established feel listened to and respected. Rapport-building techniques include using the interviewee's name, appearing relaxed and friendly, and ensuring the interviewee that there is plenty of time to listen.

The interviewer must recognize the difficulties of talking about the crime, especially a sexual crime. It is important for the interviewer to convey to the victims that he or she understands the difficulties of talking about such a sensitive topic. It may be helpful to suggest they think of the interviewer as a doctor who needs to get the details of the assault.

"Because of the nature of this incident, we'll be using sexual terms like *penis* and *vagina*. I work regularly with victims who have been in these unfortunate circumstances, so I'm comfortable discussing sexual matters. I do realize that it can be difficult to discuss personal matters with a stranger. So take your time, and if it would be helpful, think of me as a doctor who you need to talk about a medical ailment."

EXERCISE 4.2

Rapport-Building Exercises

Activity: This activity should be completed by triads composed of students who do not know each other well. Special effort should be made to assign dissimilar students by gender, race, and/or age. Each interviewee will be a victim, witness, or probation client.

Purpose: To allow students to practice rapport-building techniques.

Each student will interview one of the other members of the triad with the third member observing. The instructor gives each interview 3 minutes, after which time he or she will tell the students to stop. The observer then provides feedback. The interviewee adds feedback to which the interviewer can respond. The triad members then rotate roles. In the first role play, the interviewee is a victim. In the second role play, the interviewee is a witness, and in the third, a new probation client.

Description of roles:

Interviewer's role: Take a few minutes to think of some questions or comments that you can make to attempt to find some common areas. Observe the student you will be interviewing and try to adopt similar nonverbal behavior.

Interviewee's role: Respond naturally to the interviewer's nonverbal and verbal comments. Do not try to be difficult.

Observer's role: Locate your chair so you can see both individuals. Carefully observe the interviewer and fill out your rapport-building form. At the end of 3 minutes, the instructor will stop the interview or ask the observer to do so. Ask the interviewer what his or her impression was: How do you feel you did? What would you do differently? Ask the interviewee for his or her impressions: How do you think the interviewer did? What did he or she do to build rapport, that helped you feel you had something in common? Then give your impressions. Be sure to include your observations from your rapport-building form.

	Observer's Comments
Did the interviewer's face appear interested?	
Did the interviewer's voice sound natural and interested?	
Did the interviewer's nonverbal language appear inviting, but not distracting?	
Was the interviewer's pace appropriate?	
What did the interviewer use to connect with the interviewee?	
How effective was the first rapport-building technique?	
How effective was the second rapport-building technique?	

EXERCISE 4.3

Mirroring

Activity: Instructor illustrates mirroring with unsuspecting student.

Purpose: To illustrate for students the effectiveness of mirroring.

1. When the students have settled to begin class, the instructor will call one student to the front of the class. The instructor will ask the student a nonthreatening question such as "What is something you enjoy doing when you are not in class?"

    The instructor will try to casually strike the same pose as the student, e.g., crossing arms in front of chest, scratch head when the student scratches, and so on. The class or student may catch on at some point and say something about what they observed.

    The instructor will ask the student about his or her comfort level. Did his or her comfort level increase as the conversation continued? What caused the student to think the instructor was mirroring the student's nonverbal cues?

2. The students can be given an assignment to try mirroring that evening on somebody with whom they are comfortable. The instructor can make it into a writing assignment or just have students report to the class. Include the following information:

    a. How did the conversation begin? What was the first pose struck?

    b. What were some of the mirroring gestures attempted?

    c. Could the students see differences in their responding subjects' comfort level? When asked by the students, did the responding subjects' comfort level improve? Was there a point in which the responding subjects were suspicious of something unusual going on?

    d. What do the students see as the benefit of mirroring?

Mirroring   Rapport is facilitated with mirroring. Interviewers should observe and listen carefully to interviewees, adopting the same sensory figures of speech and matching and mirroring their nonverbal behavior (Inbau, Reid, Buckely & Jayne, 2004). People are more comfortable with people who look, talk, and act similar to them, so the interviewer consciously mirrors the speech patterns, speed of delivery, breathing, posture, and gestures of the interviewee. It is important not to mimic, but rather to use small movements of the head or hands and matching breathing. Tone, speed, volume, and rhythm can be matched subtly.

When the interviewer mirrors an interviewee's posture, gestures, and physiology, the same emotions are created within the interviewer. Emotions such as fear, happiness, and anger are linked to behavioral and physiological responses. An athlete who has just lost a game will have slumped shoulders and a lowered head. When a person cries, the head and eyes drop, but when a person laughs, the head rises. Using the same simple behaviors will create the sense of loss and sadness.

Try practicing in normal everyday conversations. Nod slowly when the person nods. If he or she crosses arms, try the same. Listen to uses of visual, auditory, and feeling words. The section on neurolinguistics later in this chapter will discuss the importance of mirroring using the same types of sensory words as the interviewee.

The interviewer can determine the existence of rapport by changing his or her own nonverbal behaviors such as by nodding his or her head slowly, raising a hand, or crossing one leg over the other. If the interviewee does the same, then rapport has been established. Much like a dance, the interviewee and interviewer respond and mirror each other's movements with movements of their own.

Listening

Being a good listener is the most critical skill to possess to conduct a productive interview. Good listeners appreciate not only the significance of what the respondent is saying, but also the effort the respondent is making to be candid and complete. Conveying complete attention to the interviewee is an important rapport builder. People are seldom listened to so they appreciate the opportunity to express themselves. It increases their feeling of importance. On the other hand, if respondents sense a lack of interest, they will retreat to abbreviated answers.

In addition, only when the interviewer gives the interviewee full attention is it possible to ensure the interviewee's responses are valid, relevant, and complete. When interviewers are distracted and thinking about other things, they are not carefully assessing the responses. Interviewers should have a mental checklist:

1. Is that response on track with the question? Or, did the interviewee stray off on another path of information?
2. Does that response sound complete? Or, do I need to ask some probes to get more information?
3. Does that response sound valid? How does it corroborate with other information? Are the interviewee's nonverbal cues consistent?

Chapter 5 will cover assessment of information in much more detail.

Difference in Hearing and Listening

Listening is much more than hearing. Hearing is passive; it is the automatic physiological process of collecting sound waves. In contrast, listening is active and occurs only with intent by the receiver. It is a high-order mental process that, before responding, includes interpretation and evaluation.

After the sound waves are captured, the mind translates the waves into thoughts that must be comprehended and interpreted. The physiological processes are complex and must compete with surrounding bodily demands.

As discussed in Chapter 1, because every human being has his or her own background, values, and belief system, the information received passes through a psychological filter before the interpretation is completed. Like light waves, the message becomes bent and is never received as it is sent. The interviewer always should be sensitive to the possibilities of misunderstanding and being misunderstood, leading to the need for feedback.

As interviewers actively listen, the sounds they are hearing are translated into thoughts, interpreted, and then evaluated for validity, completeness, and relevancy. There is no time to be distracted about personal or other professional matters.

Obstacles to Listening

Listening is difficult. The average person talks at a rate of about 125 words per minute. We can understand about 300 to 500 words per minute. So, there is a considerable amount of dead time during which a listener's mind can become distracted. Added obstacles include physical distractors such as outside noise, phones, other people, and interviewer fatigue.

Obstacles work both directions. The interviewer may face semantic obstacles such as the interviewee's vocabulary, accent, different meanings for the same word, or special meanings within a subculture. Slang or street talk can change the meaning of a statement completely. If the interviewer uses professional jargon, acronyms, or other professional language shortcuts, the interviewee will be confused and may answer incorrectly and become frustrated. For example, a probation officer asks a client the following question:

"I notice on your P-22 that you were 10-22. The PO Board isn't going to like that. How do you explain it?"

The interviewer or interviewee may lack interest in the area of discussion or question the need for the interview. The interviewer must work to keep the interviewee motivated. In addition, the interviewer must work hard to stay focused on the interview, no matter what else is occurring.

Listeners have a tendency to interpret any ambiguous or unclear response by bending its meaning to what they expect to hear instead of probing to get a clearer explanation. Interviewers must listen carefully, and if unsure of what is being said, they must be willing to probe for clarity. Probing will be covered in more detail in Chapter 5.

Interviewees react to emotionally charged words such as *steal, lie, sexual assault, rape, murder, mentally ill, domestic violence,* and *abuse.* Although it is important to communicate clearly and concisely, interviewers should be aware of the impact of certain words that impede listening and responding. In addition, interviewers need to possess sufficient self-awareness to be aware of their own reactions to certain words.

Attending Behavior

To actively listen, the interviewer must communicate intent to listen and pay attention. To properly attend to the interviewee and what is being said, the interviewer should physically display certain nonverbal behaviors and listen to content and related feelings. A helpful acronym to help remember attending nonverbal behavior is SOLER:

S: Squarely face each other
O: Open posture (uncrossed arms and legs)
L: Leaning toward the other
E: Maintaining good eye contact
R: Relative relaxation

To physically attend, the interviewer faces squarely with his or her body turned toward the interviewee. The interviewer's shoulders will be directly lined up with the interviewee. Open posture is mainly uncrossed arms and legs. The interviewer's hands are relaxed and often laying in his or her lap. The interviewer's body should lean slightly toward the interviewee, projecting interest in what is being said.

Good eye contact is difficult and varies based on gender of parties and the topic discussed. If eye contact lasts more than 2 to 3 seconds, the interviewee might think the interviewer is seeking intimacy or attempting to overpower. Instead, eye contact should be used to indicate sustained interest and sympathetic understanding without suggesting an intimate or aggressive relationship.

The interviewer should remain quiet and relaxed rather than exhibit a great deal of movement. If the interviewer is comfortable, his or her body and movements will not be rigid nor engage in distracting movements and sounds. He or she is aware that his or her body silently communicates and what it communicates.

EXERCISE 4.4

Using SOLER

Activity: Student dyads first use SOLER without conversation and then with conversation.

Purpose: To emphasize the importance of effective eye contact.

1. Students will pair up and demonstrate SOLER without conversation. The instructor only needs to let this part of the activity go for about a minute. What happens after a few seconds? Usually both parties begin to laugh or turn away. Why? Discuss with them the nonverbal language related to eye contact. Were there differences in comfort level if the students were different genders?
2. Students continue practicing SOLER but are requested to carry on a conversation with their partner on any subject. The exercise needs only to last 2 to 3 minutes. The instructor stops them with "At what part of your partner's face are you looking?" If the students respond with eyes, ask them for more detail: where did they actually look when they made eye contact—between the eyes, right eye, left eye, the nose? Did it matter if the dyad was composed of a female and male or the same sex?

3. As an alternative exercise, the instructor splits the class in two groups. Out of hearing range, the instructor tells one group they will be asked to carry on a conversation with another student in the class. During the conversation, their nonverbal behavior should be the opposite of SOLER: do not make eye contact, but do lean away, become distracted by their fingernails, a fly on the wall, and so forth. The students then are paired up and act out the instructions for a very short period of time (students have been known to get angry and storm out of the classroom). The instructor then discusses the exercise with the class:
   a. How did the student who was ignored feel?
   b. Did the non-SOLER behavior change the communication? In what way?

Nonverbal Communication

Nonverbal communication takes the form of auditory and visual cues. Auditory cues, or paralinguistics, are dynamic features such as tone of voice, speed, and volume. Visual cues include facial expressions, eye contact, posture, and use of arms, hands, legs, and feet. The interviewer needs to be aware how he or she is using these nonverbal cues and how to interpret the nonverbal cues of the interviewee.

Interviewer's Nonverbal Communication

Interviewers must be aware of their feelings and attitudes and how they are communicated by nonverbal communication. Much of the interviewer's delivery centers on the auditory and visual nonverbal cues.

The interviewer's facial expressions should show interest, but also a nonjudgmental attitude. If dealing with a subject who may have committed a heinous crime, it might be necessary to hide real reactions. Interviewers need to detach, acknowledge humanness, and proceed calmly. They must condition themselves to be strong enough emotionally to hear anything interviewees may want to say. When the interviewees get angry and react with outbursts, interviewers need to be ready to withstand the heat and to not react in a defensive way or show extraneous facial movement.

The interviewer will need to adjust to the mood of the interviewee so the interview appears to be spontaneously conversational and avoids any hint of a routine, bureaucratized reading of a script. Just the memory of standing in line at a state department of motor vehicles and experiencing the bored, indifferent interaction of one of its clerks should make every interviewer become animated and swear off any chance of becoming a similar bureaucratic robot.

The interviewer should help direct the respondent's behavior. So if the interviewee appears to be agitated and speaks rapidly, the interviewer uses a soothing tone. If the interviewee needs to be roused from a depressed, tired mood and to be stimulated, then the interviewer will sound more animated.

The rate of speech or frequency of turn taking during the dialogue also impacts the validity and completeness of the information obtained from the interviewee. The interviewer may talk faster when asking routine questions such as name, age, and date of birth, but will talk slower when asking the interviewee to recall events or search emotional ambivalences. Talking slower, the interviewer eliminates the perception of time constraints and conveys that the interviewee's information is valuable.

The interaction pattern between the interviewer and interviewee is established early. If they start at a slow pace, it will be more conducive to thoughtful interaction. If they start quickly, it may convey a lack of attention to thought or feelings of anxiety. The interviewer should match his or her speed with the interviewee's speed. The interviewer should restrain a desire to talk too quickly; it is better to slow down and have time to concentrate, think, and plan. The interviewer also is communicating less stress and anxiety if talking slower.

Use of Silence   Human beings have difficulties with gaps in talking. We have a tendency to leap into conversation at the first pause without waiting to see if the other individual intends to say more. More than likely, the other individual would have elaborated if the silence had not been broken. Silence acts as a comma at the end of a statement—time to breathe. It also communicates a mood. The lack of pauses indicates that the interviewer is anxious and insecure. The best use of silence includes good attending behavior—leaning forward and looking expectantly, but relaxed. These periods might be up to 10 seconds if the interviewee seems to be collecting their thoughts, or a delay of at least 2 seconds as a turn-taking gesture. The interviewer should not interrupt or finish sentences and should consider using minimal encouragers, such as uh uh, I see, and what else.

The goal of interviews usually is to gather information from the interviewee, so the interviewee should do most of the talking, although in the direction led by the interviewer. Silences and long productive pauses will help establish such a pattern. Consider the following examples (*y* is the interviewer, and *x* is the interviewee. Each letter designates a statement by the interviewer or interviewee.):

Yyyyxyyy_ (silence) _____yyxy. With this example, the interviewer has established that he or she will do most of the talking and will break silences.

Yxxxxxyxx _____ (silence) _____xxxy. This example demonstrates an encouragement for the interviewee to talk the majority of the time, and the interviewer has the interviewee break the silences.

Interviewee's   The nonverbal clues of interviewees can help interpret their verbal re-
Nonverbal Clues   sponses and determine their emotional state, mood, or attitude toward the topic. While words can be monitored and selected before actually being said, involuntary nonverbal behavior is not so easy to control. Verbal and

nonverbal messages are normally congruent, or in harmony, if the interviewee is comfortable. When the interviewee appears to be saying one thing, but displaying nonverbal cues that appear to be saying something else, it is worth examining further.

If the words do not match the nonverbal behavior, the nonverbal behavior is "leaked." Leakage cues are nonverbal acts that give away information that the interviewee is trying to conceal (Edelmann, 1999). If a suspect begins to pick lint off his or her clothes each time the interviewer asks about the offense, the suspect is unintentionally communicating a different message than what is being said.

However, it is not safe to assign meaning to a very specific nonverbal cue. It should not be taken out of context, and there are different kinds of influences.

It is especially important to use caution in interpreting nonverbal communication in relation to deception. Deception cannot be directly interpreted from nonverbal cues. Nonverbal behavior indicates emotions, and emotions might indicate deceit. So, first the emotional interpretation has to be accurate, and then a precise connection for the reason for the emotion must be ascertained. People overall are not accurate in judging whether others are telling the truth. The research has found that when attempting to detect deception while observing nonverbal cues, law enforcement professionals' rate of accuracy is no better than chance (Edelmann, 1999). For example, some truthful, socially anxious people work hard to make a particular impression. Through their increased anxiety, they may display nonverbal behavior that appears deceitful.

There are also cultural differences to the extent that emotional expression is encouraged or suppressed. There are different meanings to specific gestures. These differences are not only among countries but also rural–urban differences, social class differences, ethnic differences, and age.

Women and men engage in different nonverbal behavior. Men use more leg and foot movement when telling the truth, while women use more facial expressions. Women use more qualifiers (e.g., most of the time, often) than men when telling the truth. These behavior frequencies reverse when men and women are lying. In other words, men suppress leg and foot movement, and women suppress facial expressions and qualifiers when attempting to deceive (Fatt, 1998).

Different situations call for different nonverbal cues. An individual feeling in danger will desire more space than friends on the street shaking hands. In general, during interviews associated with offenses, there is anxiety. It could be associated with fear that the truthful message will not be believed or with the desire to lie.

Observing individuals while gathering background information and during the rapport-establishing phase will serve as a reference point. The interviewer can gauge the amount of space that an interviewee needs to feel comfortable and the level of nonverbal expressiveness used by the interviewee.

Cautions in evaluating behavior should expand to the surrounding environment. Distractions may cause behavior unrelated to the question asked, e.g., eye contact broken because somebody walks by. The individual who does not like law enforcement will give negative feedback. The low intelligence of the suspect may delay his or her response rather than indicate an attempt at deception. Higher educated–higher intelligence individuals may use less nonverbal gestures and be more verbal. Medical conditions or drug–alcohol use lead to physiological changes, such as dry mouth and tremors. If intoxicated, the individual is unsuitable for behavior assessment.

Change is what is important. Does a certain behavior only happen at certain times when the interviewee is responding? For example, adjusting glasses may be more than just resettling the glasses, but rather screening the face and eyes and giving time to think of a response. Does the interviewee pick imaginary lint off his or her shirt before answering questions?

Robert Edelmann (1999) wrote an extensive review of the research that studied individuals' ability to correctly detect deception and the factors that contributed to correct detection. Several researchers substantiated that there were certain behaviors that reliably distinguished deceptive from truthful interviewees, but there must be a period of baseline behavior. For instance, the judges of the behavior needed to see the behavior of the interviewees while they were answering information that could be substantiated. Again, did the interviewee's behavior change between responding to routine questions and answering critical questions?

In observing possible change in baseline behavior, the length of the response becomes important (Fatt, 1998). Answers to close-ended questions are brief so there is less opportunity to leak cues. If asked open-ended questions, there is more opportunity to observe the interviewee and the potential for more avenues of leakage (facial expressions, tone of voice, body language).

Physiological Reactions to Stress

Detecting changes in nonverbal communication when an individual is feeling anxiety is based on the assumption that people tell lies less frequently than they tell the truth, so there is a tendency to feel less confident and more anxious when attempting to deceive. These emotions translate into stressful reactions. When nervous, the individual's autonomic nervous system secretes adrenaline, which increases the heart rate and the need for more oxygen, so there is a more rapid respiratory pattern.

There is tension in the upper chest muscles so the individual cannot take in enough oxygen and may begin to pant, or may instead have slow, shallow, or irregular respiration. Irregularity may become highlighted with deep breaths periodically taken during the course of the interview and released as deep sighs. The body begins to build up heat so there is a need to dissipate heat (for further explanation, see Chapter 12). To cool, the body diverts blood from the digestive track to the surface of the body, which leads to perspiring.

Pulsation of the carotid artery on either side of neck or blood vessels in the temple may be observed as there is an increase in blood pressure. The individual may scratch as blood is diverted from the digestive tract to the surface of skin, and the capillaries expand and take additional volumes of blood, causing tingling.

The mouths of stressed subjects often are dry. The saliva takes on a tacky, stringy appearance and causes the tongue to stick to the roof of the mouth, so the individual makes a dry clicking sound and may continually lick lips.

This response is indicative of stress, not necessarily untruthfulness; however, untruthfulness is stressful to most individuals. While individuals can control some nonverbal behaviors, actions related to physiological reactions are often uncontrollable. Edelmann (1999) describes a *leakage hierarchy*. Individuals are most likely to be able to control the verbal content the most, followed by facial expressions. Body activity and tone of voice are at the lowest end of the hierarchy as the less able to be controlled. Zuckerman, Koestner, and Colella (1985, as cited in Fatt, 1998) go so far as to state that facial expressions are so controllable that nothing can be learned from them alone. It is only when other nonverbal behavior is added that facial expressions can help discern discrepancies between the verbal content and nonverbal cues.

Eye Behavior

Normally when making eye contact, it is an indication of a willingness to communicate. Lay people assume that people who are lying will avoid eye contact; however, professionals who study deception find that people who are attempting to lie will increase their eye contact. It is possible that deceptive interviewees know they are supposed to make continual eye contact if they are to be believed.

Several variables affect the amount of eye contact. There usually is less contact when physically close, discussing difficult or intimate topics, being disinterested in the topic, feeling ashamed or hiding something, or being of a cultural origin that discourages direct eye contact. The interviewer needs to be sure his or her own behavior, such as sitting too close or exhibiting too intense eye contact, is not affecting the individual's behavior.

Variations in eye behavior within the context of a particular interview furnish clues, especially if accompanied by frowning or squinting. When asked questions, most people pause, look down or up, and then look back before answering. They are thinking through the meaning of the questions, recalling events, and/or trying to put thoughts into words. If they freeze and gaze at an object or into space, it may be an indication of fear or anxiety.

Experienced interviewers may sit close enough to observe changes in the pupil of the eyes. Dilation is momentary, so it is difficult to observe. If the pupils dilate, it is a physiological indication of arousal that occurs with anxiety, excitement, interest, or satisfaction. Accompanying pupil dilation are increased eye blinks in which the eye lids close quickly and frequently over the eye.

Mouth

Humans are capable of expressing more than one emotion at a time, e.g. smiling mouth–sad eyes; however, these multiple expressions often are incongruent with each other and what is being said. Also, the mouth is more restricted than the eyes because it is used for speaking. So expressions using the mouth are most clearly seen during pauses in speaking. Pressing the lips together may show stress, determination, anger, or hostility; quivering of the lips may suggest anxiety; and surprise or shock may be indicated by an open mouth. The mouth is given such jobs as chewing on a fingernail or picking teeth when individuals are attempting to obscure their expressions or delay a response.

In a meta-analysis of 19 studies (Zuckerman, DePaulo & Rosenthal, 1981, as cited by Edelmann, 1999), the researchers concluded that individuals attempting to deceive smiled slightly less often. There also is a difference in truthful and deceptive smiles. Smiles of interviewees who are actually experiencing enjoyment include muscle activity around the orbit of the eyes as well as the muscles around the mouth. When enjoyment is feigned and the smile masks strong negative emotions, there are traces of muscular activity associated with the negative emotions. They are miserable smiles.

Different types of nonverbal cues are interconnected and congruent in manifesting the same attitude. For example, when one smiles, the ends of their lips curl up, and their eyes brighten. Without the change in the eyes, observers will not consider the smile genuine.

Head and Shoulders

The head is used to transmit *emblems*, nonverbal gestures that are directly translated into words, such as nodding or shaking. Close observation may reveal the head momentarily moving in the different direction from the answer that the interviewee verbally says. Paul Ekman, who has been conducting research in the area of deception for more than 30 years, provides expert consultation for a television series called *Lie to Me.* In one of the episodes, the heroes declare that the person is lying. They then show the videotape of the person responding. The person's head slightly nods up and down before quickly shaking side to side as he emphatically declares, "No, I was not involved." While such observations make a great television program, if such changes actually occur, they are microscopic and are not likely to be observed naturally.

To reduce tension, individuals may roll their head or move their shoulders back and forth. Tense muscles often make the entire upper body move together rather than just the head. The movement appears stiff.

As tension and anxiety develop, the shoulders rise. While relaxed but attentive, shoulders are lowered in what is considered a normal position. To know what is normal, the interviewer must know what is normal for that person.

Arms and Hands

Individuals use their arms and hands to express themselves in voluntary and involuntary ways. When animated and relaxed, there is an energetic fluid movement of hands and arms. These movements support, emphasize, or illustrate what is being said. When individuals are anxious, their muscles tense resulting in jerky, abrupt, and inappropriately timed movements. Fine motor coordination is lost, so interviewees' handwriting may change. Their signature may become illegible or have a spiky appearance.

Edelmann (1999; also see Fatt, 1998) notes that hand movement decreases with attempts at deception because telling lies takes thought and is cognitively more complex than telling fact. There is no need to be creative with the truth. The deceptive interviewee must focus on creating (and remembering) the lie so he or she will have less ability to focus attention on maintaining accompanying bodily movement.

Certain gestures are instinctively protective, as if protecting vital organs. For example, if somebody is perceived to be too close, individuals rotate their trunks, turning one shoulder away, thus protecting the vulnerable portions of their body. They also use their arms and legs for protection. If feeling safe and eager, individuals lean forward attentively, moving toward the other individual.

Although crossed arms can indicate the individual is cold, especially if hands are flattened and placed under the armpits and the trunk of the body hugged, crossed arms often suggests reservation, avoidance, or dislike of the interviewer or topic. The crossed arms can be modified so that rather than a full arm-crossed barrier, the hand may come across to touch a watch or ring on the other hand, providing a barrier and reducing stress. As already noted, the interviewer needs to be observant and assess whether such behavior happens when certain information is discussed or a specific question is asked. If the interviewee just sits with crossed arms and that motion is static, it is not possible to discern that crossed arms mean anything.

Unless interviewees carefully keep their hands in their lap or sit on them, they will usually put their hands to use. Individuals' hands may naturally tremble from illness, but trembling can derive from repressed anger or anxiety.

Individuals may tap or drum their fingers to overtly communicate boredom, impatience, and even frustration. Steepling fingers may be an attempt to appear confident or intelligent. The interviewee may use the gesture to intimidate. It is not a useful gesture for the interviewer to use.

Individuals may use their hands to cover their mouth. This action usually is an unconscious gesture to stop their mouths from saying inappropriate statements, or they may believe they have bad breath or bad teeth.

They may use their hands to screen their eyes if they are uncomfortable looking at the interviewer or having the interviewer look into their eyes. Covering their eyes may be accompanied with slumping forward with their arm propped on their leg.

Grooming gestures, such as picking lint or hairs off his or her clothes, does not necessarily mean that an individual is anxious or deceitful; however, it may be used to cover uncertainty when asked a question he or she is not prepared to answer. The interviewee is asked a question, and grooming is used to provide a pause before answering.

As a show of attitude, individuals, especially young males, will tuck their hands under their armpits with thumbs extended upwards. This pose often is accompanied with crossed legs and feet sprawled out and extended to show they are not intimidated, and possibly hostile.

Legs and Feet

Although legs and feet usually are less expressive, they often are worth observing because people do not think about attempting to control them. Also, at an increased level of anxiety, the individual will leak, or express nonverbal cues, no matter how hard he or she attempts to control them. Usually there is little leg or feet activity if the interviewee is comfortable and relaxed.

Crossing and uncrossing legs repeatedly is an area of instinctive protection that may signal anxiety. There are three basic positions: cross at ankles, knee over knee, or ankle on knee. As a general rule, the more anxious or defensive an individual becomes, the higher the knee rises to protect the abdominal region. It might even include a draped arm. The interviewee may have knee over knee with a foot directly pointed at the interviewer. This position keeps the interviewer at a distance and reduces stress. The interviewer should look for changing positions during significant questions.

The interviewee may take a runner's starting block position. Much like a runner getting ready to leap forward in a race, the interviewee will be leaning toward the door with hands on the arms of chair. Also, tapping or circling a foot may indicate stress or impatience.

Paralingual Cues

Voice quality or tone and other sounds associated with talking are other areas that leak information and are low on the ability to control on the leakage hierarchy. Individuals who are deceptive pitch their responses higher and with more hesitant answers than in their normal speech. Once again, the interviewee's baseline paralinguistics are important to gauge the person's normal pitch and rate of speech.

Paul Ekman and colleagues (1996; 1999) have concluded that the behavior of highly motivated deceivers differs significantly from those who are less motivated. A large number of studies about deception were originally conducted with students in lab settings. These safe environments did not provide sufficient motivation, or "high stakes." When observing people who are highly motivated to get away with their lies, they actually become more obvious. This *motivational impairment effect* concludes that the harder deceivers try to suppress or control their behavior, the more obvious their change in behavior. Their responses become slower, higher pitched, associated with more speech disturbances (ahs, uhs) and accompanied by less eye contact, less eye blinking, and fewer head movements.

The exception to the motivational impairment effect are cases in which the deceivers are experienced liars, have time to plan and practice their responses, and are confident in their ability to deceive. In a study of experienced sales people in which they were asked to make sales pitches for products they liked and those products they did not like, observers were not able to distinguish between their feelings (DePaulo & DePaulo, 1989, as cited in Edelman, 1999). The sales persons were experienced and had confidence in their abilities.

Neurolinguistics: Language of the Mind

"You have given me a pretty clear picture of the man's appearance. As you lay on the ground and watched him run away, did you see anything else?"

Neurolinguistic programming (NLP) is built on the theory that individuals encode information using a dominant sense such as vision, hearing, or touch. If an interviewer can decipher the dominant sense of the interviewee, then rapport can be built easily; communication differences are bridged, and more information is likely to be collected (O'Connor & Seymour, 1990). NLP also has expanded to facilitate the assessment of the validity of the interviewee's statement.

Although individuals process information through all their senses, they tend to experience in one dominant sense, usually sight or hearing. A minority of individuals may be kinesthetic or touch-and-balance dominant. Taste and smell are not used as often, especially in Western culture. Many people primarily see an event or situation; others will absorb the event through what they hear. Then, they use those same primary senses to retain those experiences in their minds. For example, when a crowd of people are experiencing a parade, many of them will use sight as their primary sense to absorb the instruments in the bands, the colorful uniforms, and the vibrant floats. Other people, while seeing the same instruments, uniforms, and floats, will focus more on the music, cheering, honking, and other sounds. When asked about the parade, those people whose dominant sense is sight will describe what they saw. Those whose dominant sense is hearing will describe what they heard. People's senses are outwardly used to comprehend the world and inwardly to represent the experiences in their minds.

EXERCISE 4.5

Neurolinguistics Exercises

Activity: Individual activity for students in which they reflect a past experience.

Purpose: To facilitate the students' understanding of how people input stimuli.

The students should think of their last birthday, last vacation, or an identified special event. Then, the instructor should ask them to describe it. As each student describes the event, the rest of the students should write down words they hear the presenter use to describe the sensory stimuli. For example, are they visual, auditory, or feeling words?

A variation of this exercise is for the instructor to divide the class. Outside of the hearing range of the other half of the class, the instructor tells half the class they will be listening and writing down sensory stimuli words used by the other students. The other half of the students will be given the instruction to think about their last birthday, last vacation, or an identified special event. Then, the instructor should ask them to describe it. The students instructed by the teacher will write down the words used to describe the events.

Building Rapport When making an effort to establish rapport, communicating in the same sensory language facilitates the process. Often, how words are communicated is more important than what is said.

For example, "I see what you are showing me," and "Try to look back and see if you can recall" can be used for visual-dominant subjects. If the witness says, "This doesn't ring true to me" or "I don't like the sound of this," consider asking "What did you hear next?" or "What did he say?" (auditory-dominant subject). If he or she says, "Let me get a grasp of this situation," phrase questions such as "How did you feel when you saw this happen? How do you think they felt?" (kinesis-dominant subject). **Table 4.1** lists samples of words and phrases for each of the three primary senses.

Respiratory patterns often are reflected in the voice, so those interviewees who are visual-dominant generally speak quickly and with high-pitched tones.

Table 4.1 Words for the Three Primary Senses

Visual	Auditory	Kinesthetic
Beyond a shadow of a doubt	Clear as a bell	Get in touch
Pretty as a picture	Keep your ears open	Have a feel for
Farsighted	I hear you	Chip off the old block
Blind to	Rings true	Get a handle on
In the dark	Give a hoot	Sharp as a tack
Seeing red	In tune with	Stiff upper lip
It is clear that	Voice an opinion	If it feels right

Their breathing is higher and shallower in their chest than other sensory-dominant interviewees. Those who are auditory-dominant often have rhythmic, expressive, and clear voices that resonate. They breathe evenly over their entire chest area. Kinesis-dominant voices are slow and have a deep quality. They are characterized by deep breathing, low in the stomach area (Zulawski & Wicklander, 1993).

Individuals can be shifted from one sensory operation to another. Certain things such as a song or a flash of light will bring up certain sensations that are re-experienced. The interviewer can move the individual from where he or she is to where the interviewer needs the interview to go. For example, an auditory-dominant witness says, "I tell you what. I have never heard anything like that before in my life. That guy said, 'Give me the money or I'll kill you.'" The interviewer will ask questions first that are related to what was heard. To transition to another sense, such as visual, the interviewer can say, "When you heard the shell click into the chamber, what did you see?" If the witness is kinesis-dominant, the interviewer can say "When you were so frightened, what did you see?"

It is most constructive to begin with the sense with which the witness associates before moving to another sense. The interviewer wants the witness to become comfortable, and talking in the most comfortable sense will help.

Use of Lateral Eye Movement

The use of the three different sensory channels can be observed by corresponding lateral eye movements (LEM) and positions for each sense. These movements occur quickly and can be easily missed (O'Connor & Seymour, 1990).

For visual memory, the eyes move upward, to the interviewee's left, and straight ahead. It is not that he or she will find the information in those locations, but rather the movement of the eyes to that spatial quadrant allows the mind to recover visually stored data.

For visual construction in which the interviewee is building an image rather than just retrieving a memory, he or she constructs an image, evaluates it (looks at it), and then answers. The visually oriented person would look upward and then to his or her right to construct.

For auditory memory, both eyes go across and to the left as the interviewee remembers external sounds. With auditory construction, the eyes go across and to the speaker's right. When accessing feelings, the eyes go down and to the left.

Left-handed people often reverse the eye movement. If they reverse them, the reversal is consistent for all three major senses.

According to Zulawski and Wicklander (1993), moving eyes left and down indicates an internal dialogue within the interviewee. Internal dialogue is represented by the head tilted forward and down with eyes down to the left. In social situations, when asked to respond to an emotional situation, the eyes may go down and to the left as conflicting thoughts are debated and the appropriate response is considered.

EXERCISE 4.6

More Neurolinguistics Exercises

Activity: In dyads, students will observe their partner's lateral eye movement (LEM).

Purpose: To demonstrate to students that LEM exists and to provide them an opportunity to practice.

The instructor will divide the class in half. One half of the students are secretly told to watch their partner's eye movement after asking each question or given each request. The students observing should make an abbreviated note after each observation. For optimum practice, the instructor presents the questions–requests on an overhead slide. The students who are responding to the questions–requests should be sitting so their backs are to the slide before the instructor presents it. The half of the students who will be asking the questions–making the requests will be facing the slide.

After the students have been asked all of the questions from the different sensory areas, the observing students will report what they saw. The instructor can write the observing students' answers on the board. Discussion should include the direction of the eyes, but then also any other eye movement. It is useful to include a note about whether the students responding are right handed or left handed. Many left-handed people will reverse the LEM directions.

The instructor should pick a minimum of one from each sensory memory.

Visual memories:

1. What color is your bathroom?
2. What was the color of the car you first learned to drive?
3. What do you see on your journey to the nearest shop?
4. How do the stripes go round a zebra's body?
5. Which of your friends has the curliest hair?

Visual construction:

1. What would your bedroom look like with purple walls?
2. If a map is upside down, which direction is southeast?
3. What would the offspring of an elephant and zebra look like?
4. Imagine a lavender circle inside a burgundy square.
5. How do you spell your first name backwards?

Auditory memory:

1. Who is the first person who spoke to you this morning?
2. Listen to your favorite piece of music in your mind.
3. Is the third note in the national anthem higher or lower than the second note?
4. What is the sound of your cell phone ring?

Auditory construction:

1. How loud would it be if 20 people shouted at once?
2. What would your voice sound like after inhaling helium?
3. Think of your favorite tune played at double speed.
4. What would a chainsaw sound like cutting through a corrugated iron shed?

Internal dialogue:

1. Recite a nursery rhyme silently.
2. When you talk to yourself, where does the sound come from?
3. What do you say to yourself when things go wrong?

Kinesthetic sense:

1. How would it feel to sit in a tub of warm Jell-O?
2. What is it like to put your toes into a mountain stream?
3. What is it like to feel wool next to your skin?
4. Which is warmer now, your left hand or your right hand?
5. How do you feel after a good meal?
6. Think of the smell of ammonia.
7. What is it like to taste a spoonful of very salty soup?

Physical Settings

Although the interviewer may not always have control over the physical setting of the interview, it is an important factor and can play a critical role in the interviewee's willingness to talk openly and candidly. The setting includes the area in which the interview will take place, what the setting contains, the seating arrangement, and environmental elements.

Interview Area Privacy is mandatory in the area to be used for interviewing. Ideally, the interview area is a comfortable room in which lack of disruption can be ensured. The interviewer's office works well because it usually contains items that personalize the interviewer. Some interviewers add props, such as a poster of the home football team or a family picture with pets, which they believe will interest and allow connection with interviewees. They also want to include amenities such as tissues and nearby toilet facilities. It is important that the telephone and other devices that might cause disruptions be turned off, and a policy is in place that will prevent interruptions.

At times, the interviewer will need to interview in the field and will have little control over the setting; however, as much privacy as possible should be requested in a confident voice. For example, at a home visit a probation officer should state that other family members need to leave the room unless there is a specific reason to include them in the interview. At a crime scene, an officer should direct a witness or victim to move to an area that is quiet and where conversation cannot be overheard by others.

Seating Arrangement   Before considering the placement of chairs, the interviewer wants to make sure there are no physical barriers such as desks between them. If the interviewing space includes a desk, the interviewer should situate the chairs so the interviewer's chair is at the corner of the desk. If notes are to be taken, the corner can be used. For safety, the interviewer always needs to be aware of the exit, and when possible, to be closer than the interviewee to it.

Normally, the chairs should be of the same type and height, facing each other, and conversational distance apart. This distance varies by country, culture, gender, the relationship of the interviewee to the interviewer, and the situation for which the interview is taking place.

Researchers in the area of *proxemics*, the study of spatial distances that individuals retain between themselves and others, conclude that we all have an invisible boundary around us that narrows or widens depending on who we are talking with. Most people in the United States have intimate conversations about 18 inches apart with this space widening to 3 or 4 feet for more impersonal transactions. Individuals feel freer talking about different topics at different distances.

Often in situations when individuals are uncomfortable or feel threatened, they will widen their personal space to outside arm's length or around 6 feet. Although sitting closer allows for softer voices and the observation of finer nuances of vocal and facial expressions, the interviewer needs to be observant for signs of sitting too close. These signs include interviewees leaning back, turning sideways, looking away, or folding their arms.

The angle of the chairs as the interviewer and interviewee face each other should be considered based on the type of interview. Facing directly in front of each other may appear threatening, but it also allows the interviewer to make good eye contact and show open posture as discussed earlier. The interviewer may want to slightly angle the chairs to decrease the potential of the interviewee feeling uncomfortable.

Other Physical Setting Concerns   The lighting of the room and note taking–recording are two other concerns that need to be considered. Lighting should allow each person to see subtle changes without harshness of a glaring light that shines directly into the interviewee's eyes.

In most cases, note taking and/or recording should be conducted without being too obvious but without hiding the activity. The interviewer should set up the tape recording before the interview and ensure the tape is sufficiently long to last the entire interview. The interviewer should be educated in the legalities of taping, especially in the area of notification to the interviewee.

Verbal Setting

As noted earlier, the first few minutes of an interview establishes impressions that will influence the interview. The interviewer needs to plan what he or she is going to say during the introduction. The introductory minutes include identification, rapport building, and anything that will reduce the interviewee's inhibitions and increase motivation to talk. The interviewer must take this time to establish rapport and set the tone for the remainder of the interview.

The interviewer should make first impressions count because the interviewee starts assessing immediately in making the decision of whether it is safe to speak. If the interviewer uses a friendly tone for voice and welcoming eye contact and has a calm demeanor, the interviewee is more likely to feel respected and at ease.

At the same time, the interviewer can begin processing information from the interviewee during the introduction. The interviewer should shake the interviewee's hand and evaluate the individual's handshake. Does it tremble? Is there reluctance to extend a hand? Does the hand feel warm and dry or cold and clammy? In this way, the interviewer can begin to establish a baseline of behaviors to observe. Even from the beginning of the interview, the interviewer may note evasiveness and lack of cooperation. The interviewer processes what is observed and tries to determine if the interviewee is signaling nervousness, fear, or deception.

During these first few minutes, the interviewee may take the opportunity to vent emotional energy. He or she may be irritated that he or she had to take time to drive and meet with the interviewer. Within reason, allow the venting and respond in a way that the interviewee feels heard:

> "I know it was a hassle to drive downtown and find a parking place. I can tell as a citizen you realize how important it is for you to help us. We really appreciate your willingness to take this time. I understand you may have information that pertains to our investigation."

As part of setting the tone, the interviewer will announce the objectives of the interview, thus answering the interviewee's unasked question of why he or she is there:

> "I want to determine how the incident we are investigating happened. With your help, we hope to prevent similar events. You are one of several people I am interviewing, so I want your assistance to get a better view of the circumstances."

These statements show appreciation for the interviewee and invite him or her to share any thoughts, observations, opinions, or facts.

Raymond Gorden (1998) suggests several questions that the interviewer should consider to facilitate a motivated interviewee.

Who is this interviewer? Some thought needs to be given to the introduction. The interviewer needs to be seen as credible and nonthreatening.

Quite often, criminal justice professionals become efficient bureaucrats who want to deal as quickly as possible with the next person. Call it indifference or haste; nothing inhibits gathering information quicker.

What can you do for me? If the interviewee initiated the interview, then the interviewer can ask the interviewee directly what he or she wants and respond with what can and cannot be done. In most cases, the interviewer as a criminal justice professional initiated the interview, so the interviewer will provide a mutual beneficiary comment, e.g., "I know you want to see justice done," or "I want to ask you some questions so you can have a smooth probationary period."

Why do you want to interview me? Briefly outlining the objectives of the interview at the beginning in a nonthreatening and anticipatory–motivating way will help the interviewee feel more comfortable and less guarded.

Why are certain types of personal information needed? The interviewer should anticipate what information may be considered sensitive to the interviewee and be prepared to explain why specific information will be needed. For victims, it also is a way to prepare them.

> "Jane, in order to understand your attacker and find him, I will need as much information as possible, and what you know is critical. At times during this interview, I will need to ask you some questions that are personal and may make you uncomfortable. Please take your time, and I will try to find a way to ask them that make you less uncomfortable."

Why are you taking notes and/or tape recording? The tape player or note paper should not be hidden, but also their use should not be distracting, so they should be already set in place. The interviewer might want to consider stating, "I want to use tape recorder so I won't forget important things you say; also using the tape recorder, keeps a third person from being in the room, taking notes. I will be the only one listening to the tape [if true]." The interviewer's tone is important and should sound confident and certain, not doubtful, apologetic, or anxious.

Will what I say be kept confidential? As soon as possible within the interview, the interviewer should outline exactly the parameters surrounding confidentiality. Confidentiality is different for each criminal justice agency. One important rule is do not promise more than can be ethically delivered.

Conclusion

During the planning of the interview, the interviewer will use the three phases of the interview to help gauge what will be covered, keeping in mind the differences between an investigative interview and a helping interview.

Rapport building is another important skill that needs to be utilized from the first interaction. Incorporating what the interviewer has researched about the interviewee into a way to relate and connect will relax the interviewee and elicit valid responses. Active listening serves to help develop rapport and to listen more fully, continually assessing whether the interviewee's responses are complete, relevant, and valid.

Nonverbal behavior works both ways; the interviewer communicates to the interviewee as well as the interviewee to the interviewer. Communicating interest through SOLER should be the first priority. The interviewer should be cautious in interpreting the interviewee's nonverbal behavior, but when changes in behavior are observed, further exploration may be warranted.

Finally, planning the physical setting for the interview and the introductory remarks is important. The physical setting must be welcoming and a tool to develop rapport. If the interviewer carefully plans the interview's introduction, then the rest of the interview will more likely be productive.

ALLEN'S WORLD

Because I rarely control the interview setting, I have to be creative in building rapport.

When interviewing a client who is in a jail cell, behind glass, I put my hand up to the window. Most prisoners understand this form of greeting. I am building rapport.

Depending on the case, and the client, I ask if there is anything I can do for them on the outside. Can I contact someone or help someone get someplace? I am building rapport.

One trick Dr. Lord does not mention is clothing. I don't interview many bank presidents. Most of my clients are middle class. Many of them are poor. I have found that the more casual I dress, the more comfortable a client is with the interview process. They do not want someone in a Brooks Brother suit interviewing them. It reminds them too much of FBI agents. I wear T-shirts, blue jeans, and sneakers. My hair is rarely combed. I often have stubble on my face. I am building rapport.

At the jail, the guards often want to put me in an open area. There are family members on each side of me visiting their loved ones. As Dr. Lord emphasizes, privacy is essential. As a private investigator, I have the right, and I always enforce it, to meet in a private room. I do not want interruptions. I do not want guards listening in.

I rarely promise confidentiality to a witness, or a client. The attorney I am working for must have access to the information I gather. Also, at some point, if we want to ask for a summary judgment or a plea bargain, we share our information with the assistant district attorney (ADA).

"You have a weak case," we tell the ADA. "Here's our defense. Do you really want to try this case?"

The client must know this is a possibility.

When interviewing witnesses, I always tell them they might be called to testify in open court before friends and often witnesses for the prosecution.

When interviewing outside the jail, I usually interview in others' environments so I have to find ways to control it. I will look around the office and comment on pictures, trophies, and art objects—anything to begin the interview on a friendly basis, regardless of the fact I know the interview could get ugly.

Even in these settings, I dress casually. I almost always wear a T-shirt from my alma mater, The University of Florida. I want to disarm, if possible, the subject. Nothing starts a discussion quicker than a heated talk about football, the SEC vs. The Big Ten, or the ACC.

I move chairs around to suit my needs. I do not want a desk between myself and the subject. I close drapes, turn off lights, and do whatever I can to make the environs friendlier to me. I rarely ask permission. I just do it.

I often ask for water or a Coke. Will the subject comply?

I shake hands with the witness and evaluate it. How does the palm feel? Is it a strong grip, weak, or clammy?

I might ask to tour the facility. The tour could provide information that might be useful during the interview.

My notepad is in plain view. I usually do not ask if I can tape record the interview. I just put the machine down in plain view, turn it on, and begin the conversation. I let the subject get up the nerve to ask me to turn it off. It is another way for me to keep control of the interview.

And—this would never happen in Dr. Lord's world—I often use a hidden microphone to tape record interviews. By law, I do not have to tell the subjects I am doing this. This is my interview, not theirs.

There is a bonus, a big bonus, to working in Allen's world. It is called a subpoena. Often, if people are unwilling to talk to me, I share with them the reality of certain procedures.

Some people think I am threatening them. I am not. I tell them this: "Look, you can talk to me here, now, at your convenience, privately. Or, I can get a subpoena. Then, you *must* come and talk with us at a time and place of our choosing with a stenographer present. The choice is yours."

CHAPTER

Conducting the Interview

Upon completion of this chapter, students should be able to

1. Motivate the interviewee to respond to the interviewer
2. Reflect the interviewee's content and feelings
3. Evaluate the meaning and completeness of the interviewee's responses
4. Probe effectively
5. Differentiate between conducting an investigative interview and a helping interview

Introduction

Once the interviewer has invested the time and effort to learn about the interviewee, developed the objectives for the interview, and worded questions to generate a response that meets those objectives, the critical time has arrived—conducting the interview.

To properly conduct a productive interview, the interviewer must establish rapport, properly deliver the prepared questions, attend and actively listen to the responses, and probe those responses when the interviewer decides they are not complete, relevant, or inaccurate. Interviewing is exhausting work. Interviewers cannot afford to let their minds wander.

During the Interview

Motivating Witnesses

She sat on the edge of the chair. "How could she tell this detective, this man, what had just happened to her?" she thought.

The detective leaned forward and said gently, "Ms. Green. I know you have been through a painful and scary experience. I want you to know that you are safe now. Please take a deep breath and take your time."

She briefly closed her eyes and breathed deeply, releasing some of the tension. "Now, why don't you begin from when you left work?". . .

The interviewer needs to be prepared to motivate individuals to provide information. Although in many cases people are willing to talk, the interviewer needs to have sufficient information about the interviewee prior to the interview to know what would provide an incentive. For example, has the interviewee ever been a victim before or had prior experience with criminal justice agencies?

During the first few minutes of contact, the interviewer should look for signs of a reluctance to talk, and if it is apparent, try to figure out why the interviewee is hesitant. The individual may not want to get involved, take time off from work, or testify. There may be fear of retaliation or reluctance to get somebody in trouble.

Culturally, he or she may shun any person who is involved with the criminal justice system or may believe giving information is against the interest of his or her family. These fears need to be addressed and, when possible, appropriate reassurances offered. Often there is safety in numbers. If the witness knows others are coming forward to be interviewed, his or her fears may be reduced.

> "I appreciate your willingness to talk to me about what you saw that day. A number of people have already talked to me about their observations, but I like to be thorough and cover all the bases."

If the interviewee does not have concerns, the interviewer needs to consider making a "sales presentation" that highlights the benefits of providing information. The interview process is tiring, and at some point, the witness who has nothing invested may become impatient.

These benefits can be anything that returns something to the person being interviewed, something as simple as respect and recognition. The benefit may be articulated in the form of crime prevention, such as the need to find the perpetrator to prevent other women from being raped or other people from being robbed.

The interviewer should have a positive outlook approaching interviews; all interviewees will be willing to answer questions truthfully and completely. To facilitate the interviewees' willingness, interviewers should word questions for complete and truthful responses, e.g., "When was the last time you talked to Sam?" not "Have you talked to Sam recently?"

EXERCISE 5.1

Gaining Responses

Activity: The students rephrase questions to obtain responses.

Purpose: To increase students' abilities to motivate interviewees.
    Rephrase the following questions to increase probability of a response.

1. Have you been able to stay sober this week? (To probationer)
2. Did you know that your son has been skipping school?
3. Do you have any information about the shooting that happened in your apartment's parking lot last night?
4. Can you help me figure out what it is going to take to keep you off drugs?

It is important to be patient and to resist dominating the interview. The interviewer needs to be calmly persistent, not aggressively heavy handed. If the interviewer conveys an impression that the interviewee is on the verge of reporting a significant fact, positive expectations will surround the interview.

Don Rabon, in his book *Interviewing and Interrogation*, suggests a number of additional strategies if the interviewer is faced with a reluctant witness or victim. These strategies include relating to other individuals with whom the investigator interacted. One scenario follows:

> You remind me of another woman . . . (similar situation). . . She felt miserable and didn't want to talk to anybody, much less me, a stranger. After spending some time together, she decided to share with me all that she could remember about what had happened, even though it was very painful. After recounting the entire incident, she said that she felt such relief, and if it would prevent it from happening to another person, then it was worth it (Rabon, 1992, p. 43).

Another strategy to motivate a reluctant person is to reduce a complex situation into a more concrete, understandable circumstance by using an illustration calculated to make an impression. This strategy is useful especially when the interviewee is torn between loyalties. The example should provide an explanation and a course of action. While the interviewee may have a difficult time recognizing or dealing with the conflicting emotions, she or he can understand concrete images, and it will help move him or her towards a decision. The following example relates to a reluctant victim of repeated domestic violence:

> Denise, think of it this way. You plan to drive cross-country, and you have a choice on the type of car you will drive. You can drive a sound, new automobile that will take you where you need to go safely or you can take an old car that you have had for a long while. You are comfortable with it, but it hasn't been well maintained and it has a lot of miles on it. The brakes are very safe and the battery is old so sometimes it doesn't start. Now, you have a choice. You can drive the new, safe automobile or the old, dangerous car. You are comfortable with the old car, but it is no longer safe. Which car are you going to drive across the entire country?

A third strategy Don Rabon (1992) describes is the process of placing an idea before a person in such a manner that he or she accepts the idea as his or her own. The cooperation that the interviewer wants to elicit is connected with an actual behavior such as drinking a cup of coffee or sleeping: "As you sit there and think about this situation, you begin to . . ." or "Sleep on this tonight and in the morning you will realize . . ."

Empathy   The interviewer cannot accurately reflect the interviewee's content or feelings without empathy. Empathy is the skill and ability to accurately understand

and accept another person's thoughts, feelings, and behaviors while at the same time maintaining one's own integrity. In criminal justice interviews, the interviewer strives to be empathic with what the interviewee is trying to communicate, but not with the person's conduct.

For example, a probationer describes his or her difficulties in avoiding associates who bought and used drugs with him or her. The interviewer needs to understand and even appreciate how hard it must be to no longer associate with established friends. Nevertheless, that is not the same as excusing the probationer's return to seeking out the associates and using drugs.

Empathy leads to knowledge. As noted by Egan (1975), empathy helps establish rapport and keeps the interviewer from talking too much. It keeps the focus on the interviewee, who is the primary source of information. In demonstrating empathy, the interviewer is doing more than restating what the interviewee says. He or she is responding to expressed and unexpressed feelings and content.

Reflecting and Accepting Content

"As I left the back entrance of my workplace, I was pulling the keys from my pocketbook. It's pretty dark back there so I always try to get to my car and unlock it as soon as possible. So I didn't see him until I almost ran into him. It startled me, and I screamed." She paused, as if catching her breath.

"When you left work, you prepared to get in your car as quickly as possible. Thus distracted, a man standing in the parking lot startled you. Is that correct?"

She hesitated, "Well, he wasn't exactly standing."

Because communication between two individuals is complex and easily misunderstood, the ability to reflect, or paraphrase, periodically what the interviewee is saying and feeling is essential. Whatever the circumstances for the interview are, the importance of accurate and precise information and understanding are critical in the criminal justice field. In addition, paraphrasing shows that the interviewer is listening and interested in exactly what the interviewee is saying. Often the paraphrasing leads to elaboration of the initial statement.

When paraphrasing a response, the interviewer makes a statement that should be equivalent to the statement made by the interviewee. When appropriate, the interviewer also should include the key feeling. Identifying feelings is difficult and takes practice. This area will be expanded later in the chapter. The interviewee then is invited to correct the accuracy of the paraphrase. A lead such as "What I hear you saying is . . ." or "Let me see if I have what you've been saying" can begin the paraphrase. A check for accuracy, "Is that correct?" or "Am I hearing what you've said?" allows the interviewee to provide feedback.

EXERCISE 5.2

Reflection of Content

Activity 1: The students paraphrase the verbal content of their partner.

Purpose: To increase the students' ability to actively listen.

The students should first focus on "parroting," or repeating, back to their partners what they said. If there is time for practicing the exercise several times or for an extended period of time, the students are encouraged to move from parroting exact words to paraphrasing the core thoughts of the interviewee. In a later exercise, students will be requested to add feelings. Students can remain in dyads, or the instructor may wish to add observers to help with feedback. One student will become the communicator and the other, the listener.

1. The communicator makes a statement about himself or herself, limiting it to one complex sentence. The listener repeats the substance of what the communicator says, beginning, "You said that . . ." The communicator (or observer) provides feedback to the listener on his or her accuracy.
2. The same process is repeated, but the communicator makes two complex statements.
3. The same process is repeated, but the communicator makes three complex statements.

The roles are rotated so that each student is given the opportunity to practice.

The instructor may want to summarize the exercise with a class discussion of what they learned. How difficult was the role of communicator, listener, and observer? In what role did the student learn the most? What factors seemed to cause the most inaccuracies (speed, length of response, complexity of response, so on)? The instructor should conclude with the importance of accuracy of content in listening as the first important step.

Activity 2: The students will practice listening first and then add paraphrasing the content they heard from their partners.

Student dyads are formed. Each will take turns at interviewing. The students listen actively and select appropriate places during the interview to paraphrase. The instructor should suggest a list of topics to ask about that relate to actual aspects of the students' lives:

How do you get along with your siblings–parents?
What led you to decide that criminal justice is the field you want to work in?
What were the factors that led to you choosing (university)?
How would you describe your childhood?

After the interview, the interviewing student writes a brief social history of his or her partner based on his or her responses. The interviewer gives it to the interviewee for evaluation. While the interviewer is completing the social history, the interviewee evaluates the interviewer's interviewing performance according to the following criteria:

Criteria	Yes	No
1. Eye contact was maintained without staring.		
2. Body posture was appropriate (relaxed, slightly leaning forward).		
3. He or she made me feel comfortable and relaxed.		
4. He or she seemed to be genuinely interested in me.		
5. He or she delivered questions without hesitations.		
6. He or she often asked for clarification and often paraphrased.		
7. He or she accurately reflected the content.		
8. I felt that I could tell him or her just about anything he or she asked about my personal life.		
9. Rating of his or her reported accuracy of my social history (circle one).	1 2 3 4 5 6 Most Least	

The interviewer will read his or her evaluation while the interviewee reads what the interviewer has written for the social history. Each should then discuss what they reported.

The two students switch and take the other role; the interviewer becoming the interviewee, and the interviewee becoming the interviewer (a third person can be included as an observer).

Reflection of Feelings

"I was terrified the moment I saw him. I instinctively knew that he wanted to hurt me. I tried to turn and run back to the building, but he caught my arm and twisted it. As I opened my mouth to scream, he pulled me around and put a big, filthy hand over my mouth." She sobbed and put her face in her hands. After a few minutes, she whispered, "I can't talk to you about what happened next."

The detective leaned forward, but didn't touch her. Instead he used his voice to comfort her. "I know you feared for your life. While I have never experienced what you have, I think I know how it feels to think that you will die and never see your family again."

She looked at the detective, and continued . . .

Reflection of the interviewee's feelings is important in helping interviews and will be covered in depth. Even in investigative interviews, the interviewer will be communicating empathy and rapport building so the interviewee believes his or her feelings are heard.

Really listening includes not just what the interviewee is saying, but also what it means to him or her. Just as the interviewer learns how to paraphrase the content of the interviewee's message, the feelings also need to be reflected. As the interviewer paraphrases and then asks for the interviewee's feedback on the content, the feelings are couched in tentative words to facilitate interviewee feedback. In the example above, the detective makes it clear that he realizes he has not experienced a sexual assault, but connecting through a life-threatening experience makes an impression on the victim. The detective couches his attempt in "I think."

Most people have a limited feeling vocabulary. Robert Shearer (2005) has compiled a list of words for feelings and designed a hierarchy-of-emotions word chart. The words range from mild to intense and from negative to positive with noncommittal words in the middle. If the interviewee is talking about family members or close-working associates, descriptive adjectives will be high intensity.

If the interviewer picks too high an intensity to reflect back, the interviewee may correct the feeling word but will still explore the area with the interviewer. If the chosen feeling word is not sufficiently intense, the interviewee may not correct the interviewer, but the conversation will remain on a superficial level. Shearer warns interviewers to stay away from noncommittal words. See **Table 5.1** for examples of feelings words.

Table 5.1 Example of Feeling Words

Intensity level	Positive	Noncommittal	Tension	Anger	Fear	Negative
Low	Amused Comfortable Pleased	Curious Interested	Blocked Caught Pulled	Annoyed Bothered Irritated	Apprehensive Concerned Uneasy	Bored Disappointed Resigned
Medium	Delighted Happy Hopeful	Confused Good Bad	Locked Pressured	Disgusted Harassed Resentful	Anxious Frightened Threatened Worried	Discouraged Hurt Lonely Sad
High	Excited Fulfilled Proud Terrific		Wrenched	Angry Furious Infuriated	Overwhelmed Scared Terrified	Depressed Helpless Hopeless Miserable

Source: Adapted from Shearer, R.A. (2005). *Interviewing: Theories, techniques, and practices* (5th ed.). Upper Saddle River, NJ: Prentice Hall.

EXERCISE 5.3

Reflection of Feelings Exercises

Activities: The students will expand their feeling vocabulary and then practice communicating the feelings expressed in responses.

Purpose: To expand the students' facility in expressing and identifying feelings

Activity 1: The student should read the two examples and then complete the other emotions.

Emotions can be expressed by one word, but often people use descriptive, metaphorical, or slang phrases to express feelings, especially if they are deeply affected. It also is helpful to understand what specifically brought on the feeling and what action the individual is considering.

Example 1: Thrilled
    Single word: I'm ecstatic
    Phrase: I'm on top of the world.
    Experiential statement: I think I will get the promotion.
    Behavioral statement: I feel like celebrating.

Example 2: Anger
    Single word: I'm furious.
    Phrase: I'm so mad that I see red.
    Experiential: I feel I've been walked on.
    Behavioral: I feel like telling him to shove it.

1. Anxiety
    Single word: I'm
    Phrase: I'm
    Experiential: I feel–think
    Behavioral: I feel like

2. Confusion
    Single word: I'm
    Phrase: I'm
    Experiential: I feel–think
    Behavioral: I feel like

3. Guilt
    Single word: I'm
    Phrase: I'm
    Experiential: I feel–think
    Behavioral: I feel like

4. Rejection
    Single word: I'm
    Phrase: I'm

Experiential: I feel–think
Behavioral: I feel like

5. Depression
   Single word: I'm
   Phrase: I'm
   Experiential: I feel–think
   Behavioral: I feel like

6. Shame
   Single word: I'm
   Phrase: I'm
   Experiential: I feel–think
   Behavioral: I feel like

Activity 2: The students read the paragraph and write down how the person is feeling. Then, they should briefly describe why that individual is feeling that particular way.

1. Juvenile to court counselor: "The other kids in my class don't like me, and I don't like them. They are so mean to me. They make fun of how I dress and talk behind my back. My Mom can't afford to buy me what those snobs wear. They don't have to like me, but they better stop making fun of me."
   How does this person feel?
   Because

2. Intercity adolescent, age 17 years, talking about the police: "Who the hell do they think they are, pushing us around like that? Everyone around here is just garbage to them. If one of us just looks funny at them, they arrest us. They don't live here. They're the strangers. They hate patrolling here, and we end up paying for it."
   How does this person feel?
   Because

3. Man, 55 years old, talking with a counselor of a batterer's group: "I'm a very private person. It takes me a long time to open up to people. I sure won't reveal my innermost thoughts here. I don't think this is the place for it. I hope the others aren't going to be trying to make me 'come clean' or parade my dirty laundry around. That's just not me."
   How does this person feel?
   Because

4. Domestic violence victim talking to counselor: "He tries to control me. It's a pattern; he's been controlling me since we started our relationship. He even makes me feel responsible for him hitting me."
   How does this person feel?
   Because

5. Sexual assault victim with detective: "I don't think I can talk about it here. What happened to me is too personal. You're a stranger to me, and I don't tell personal things to strangers."

How does this person feel?

Because

Evaluating Information

Another role in active listening is evaluating the information the interviewee is conveying. The information has to be evaluated for relevancy, validity, and completeness.

One of the first steps in preparing for the interview is the construction of objectives. What is the purpose of the interview? The interviewer keeps those objectives in mind when evaluating the information sent by the interviewee. Often, the interviewee will speak voluminously while evading the crux of the subject. It may or may not be intentional that the information is not relevant to the objectives of the interviewer; the interviewee may not understand the question and wants to say something interesting.

At times, interviewees may wish to avoid talking about their actions or beliefs and stray to other people or events. It is especially ego threatening for respondents to give motivations for an action or inaction. Asking *why* an individual committed a certain crime often will lead to rationalization rather than accurate information. Few people want to be considered bad or weak. The probationer who fails a drug test is unlikely to state that he or she just used drugs for pleasure or because he or she was associating with forbidden associates, but instead may describe the stress he or she is experiencing.

Interviewees often mix up the chronological order of events or confuse what they learned after an event with what they actually observed or experienced, especially after time has passed and the interviewee has learned additional information. The interviewer should be sure to clarify what the interviewee specifically observed and the order of events.

Evaluating validity or the truthfulness of the interviewee's responses must be done within the context of inconsistencies. The interviewee may be reporting rumors or hearsay rather than what he or she directly knows. If pressured for answers, interviewees may make up responses rather than appear uncooperative or feel they are disappointing the interviewer.

Inconsistencies can be between the interviewee's facts, between facts and generalizations, and between the interviewee's statement and known facts. The interviewer needs to be alert to differences between statements. Later statements made after the interviewer has built trust and helped stimulate memories are usually more valid, but these inconsistencies within the interviewee's statements need to be clarified: "Denise, I need some help understanding a couple of things. You said that this man grabbed you as you tried to run back into your workplace, but he was not standing. Can you explain a little more where he was when he grabbed you."

The interviewee may give premature generalizations early in the interview or may have a biased opinion that prevents recognition of the discrepancy.

> "The parking lot at work always gave me the creeps after dark. There wasn't enough light so I always was worried about people lurking in the shadows . . ."
> "No, I didn't want to trouble anybody to walk me out to my car after work. Other women did, but I didn't want to appear weak."

Interviewers usually are conducting some form of investigation whether it is a criminal investigation, pretrial assessment, or disposition evaluation. Other facts that the interviewer has collected should help him or her to be able to cross-check for inconsistencies between respondents' statements, the statements of others, and other facts.

Interviewers should be careful about depending on nonverbal behaviors conveyed by the interviewee. As was discussed earlier, nonverbal behaviors are different for different cultures. Nevertheless, inconsistencies between what an individual says and what he or she displays in nonverbal behaviors should be noted. For instance, a respondent shakes his head from side to side while stating "I loved Nancy. All I have ever thought about was what I could do to please her."

EXERCISE 5.4

Evaluating Validity, Relevancy, and Completeness

Activities: Students evaluate interview scenarios for validity, relevancy, and completeness.

Purpose: To allow students to practice assessing the validity, relevancy, and completeness of the interviewee's statements

Activity 1: The following scenarios can be given to students to read and discuss. As the students' skills advance, the instructor may want students to role play the scenarios and then ask for students' assessment. A final higher level activity is for students to role play extended scenarios, but the interviewees modify their interactions to increase validity, relevancy, and completeness.

With the following responses, assess the relevancy and completeness of the answers. If the answers are not relevant, only partially relevant, or incomplete, follow up with a question that will steer the interview back to the objectives or provide a more complete answer.

1. Probation officer (PO) with client (C):
   PO: Have you been able to get a job since our last meeting?
   C: Well, I borrowed a paper from one of my neighbors and checked out the want ads, but there wasn't really anything that I could do.

Then I checked with one of my friends to see if she could keep my kids if I was able to get an evening job.

2. Court counselor (CC) meeting with the parent (P) of a truant child:

CC: I understand from the school that John still is not attending classes. In fact, his teacher says he wasn't on the school bus in the morning.

P: John must be lying to me. You know I work evening shift so in order to get some sleep, I can't be awake to get him on the bus. When I get up, he's gone so I figure he is at school.

3. Police detective (D) interviewing a potential suspect (S):

D: Sam, where were you last Sunday evening?

S: Detective, I've got a curfew. I'm not supposed to leave my house any night after 6 P.M. until 6 A.M. the next morning. You can ask my probation officer. He'll tell you that is one of the conditions of my probation. He checks on me; I'm always there. Ask him.

4. Correctional officer (CO) talking to an inmate (I):

I: I get so tired of you guys always picking on me. You all are on some kind of power trip.

CO: Why do you say that we always are picking on you?

I: Look, I don't mean anything disrespectful. I guess I shouldn't come whining to you. If I have a complaint, I guess I should take it up with the inmate advocates.

Activity 2: The instructor can divide the class into triads to have an interviewer, respondent, and observer. The alternative is to conduct two 3-minute interviews in front of the class so the students who are not participating in the interviews are observers.

The interviewers are each given an open-ended question for two topics. Some examples of questions are: What effort did you put into studying for our last exam?, and What do you think about our (sport-football, soccer) team?

The respondents are instructed in which topic they should include fabricated information. The respondents should not share the information. The interviewers have 5 minutes to prepare for the interview, and the respondents take the 5 minutes to prepare their responses (truthful or fabricated).

The instructor should coach the interviewers to follow up their initial question with questions designed to obtain as much specific and concrete information as possible. They should think about in-depth questions that require describing locations, people and their relationships with other people involved, and the influences behind events.

The respondents should make their truthful answers as freely and honestly as possible, and their fabricated responses should sound truthful also.

The observers should listen and observe carefully for any cues that appear to reveal invalid responses.

At the end of 3 minutes, the interviewers and observers complete an evaluation form, describing any nonverbal cues or verbal inconsistencies.

	Fictionalized Responses	Cues of Fictionalizing	True or False
Topic A			
Topic B			

Probing

She gave a long sigh. "I think that is it, all I can remember. I just want to try and forget about it. I want to get this awful scene out of my head."

The detective smiled and said, "I understand. You want to be able to put this behind you and get on with your life. You have done a great job. I just have a couple of areas that I need to get more information about. Earlier you mentioned that he grabbed you and twisted your arm. How would you describe his body type?"

When the interviewer carefully has evaluated the interviewee's response and found that it is incomplete, irrelevant, or invalid, tools must be used to rectify this response. There are a variety of different types of probes that can be utilized. Unlike questions that are planned, probes are improvised depending on how the interviewee answers the question. Probes may be questions or statements used to clarify, elaborate, and motivate the interviewee to give additional information.

Inexperienced interviewers often fail to see the need to probe. It is easy to be overwhelmed by the speed of the delivery and great volume of information to be digested. Also, inexperienced interviewers will be no match for a skillfully evasive interviewee. The answer may seem adequate and relevant or may seem totally irrelevant and not worth probing.

At the other end of the continuum from failing to probe is probing too quickly and too often. Excessive probing might interrupt the interviewee's narrative, free-flowing pattern. Instead, it is best to establish a thoughtful pace, waiting until the individual is through and indicates verbally or nonverbally that he or she is expecting the interviewer to speak. What may seem an eternity to a green interviewer is about two seconds of silence.

Gorden (1998) describes a range of topic control for probes. Topic control is the degree to which the probe controls the topic of discussion. The least control is active silence, and the greatest control is retrospective clarification or elaboration. Each of these types will be described.

When the interviewee is making an effort to remember information and appears to be providing relevant information, active silence that invites him or her to continue talking is appropriate. The probe does not need to control the topic. Active silence includes attending behavior and expresses empathetic interest. The interviewer is leaning forward, relaxed, and nodding his or her head where appropriate. Active silence avoids interrupting the interviewee or slowing down the pace and creates a thoughtful atmosphere. It also allows the interviewer time to formulate verbal probes.

Encouragement adds words, nonverbal cues, and gestures that indicate the interviewer is listening, accepts what is said, and reinforces the interviewee to continue speaking without changing the direction of the topic. Similar to active silence, encouragement shows interest and does not interfere with the free flow of conversation. Later in the interview, the interviewer may find that the interviewee is wandering too far afield and needs stronger topic control, but especially early in the interview, some irrelevancy should be accepted in the interest of motivating and allowing for free association.

More topic control occurs with probes that elaborate and clarify. These probes can be added immediately after the interviewee completes a thought, or later when the interviewer wishes to return to an earlier point in the interview.

There are strengths and weaknesses with both strategies. Immediate elaboration occurs instantly after the thought and is asked in such a way as to receive more information: "What else do you know about this situation?" While there is some risk of interrupting the interviewee's free flow of information, immediate elaboration remains in relevant territory and does not bias the response. Immediate elaboration controls the topic by restricting the interviewee to the immediately preceding response.

Immediate clarification occurs right after the thought but specifies the kind of information that is needed. The interviewer is requesting more detail on a specific part of the time period covered in the previous response or a specific aspect of the event. There often is a need to increase the concreteness of the response, e.g. "You said that he hit you. Where did he hit you?"

The interviewer may make the decision that it is best not to interrupt the momentum or risk an ego-threatening situation. If probing is delayed, the interviewee may provide the needed information later in the interview. If the interviewer decides to delay, it is important to keep notes with specific words or phrases from the response that need elaboration or clarification.

Retrospective elaboration expands on what the interviewee said earlier: "Let's go back to when you said your father came into your bedroom. Tell me more about that." Retrospective clarification asks for more concrete information about an earlier topic: "Earlier when you were talking about the bedtime ritual that your father kept with you and your sister, you mentioned that he would take things from a nightstand next to his bed. What did he take from the nightstand?"

Summarizing is a form of a probe. It can take the form of the interviewer controlling the summary or requesting the interviewee to recapitulate. If the interviewer summarizes, the information should be broken into segments and include, periodically, a question that invites correction, e.g., "Does that sound right? What did I leave out?"

Recapitulation by the interviewee can start at the beginning or at a specific point controlled by the interviewer such as, "I'd like for you to go back to where you were first approached by your attacker and tell me everything just as you remember." Recapitulation is used before elaboration or clarification probes. It is less contrived than responses to specific questions. It also can be used to match facts from the first telling.

There are certain areas to avoid when probing. Because loaded, and especially emotionally loaded, probes communicate judgment, they can cause the interviewee to become defensive. It is best to begin with neutral, low-topic control probes, such as silence and encouragement.

For the interviewer to state, "Let's talk about your participation again. I don't think you meant what you said exactly," communicates to the interviewee that the interviewer desires a different response. To help the interviewee remember what was said, ask for elaboration or clarification using the respondent's words, but do not tell the interviewee that he or she was wrong or confused.

Probe Notes

To avoid forgetting what specific points need to be probed and remembering exact words of the interviewee, making probe notes is critical. Probe notes do not take the place of taping the interview, but rather are short words or phrases that do not take away from the ongoing interview. When possible, the interviewer should jot down exact phrases that the interviewee used.

Wasn't exactly standing—perp position?
Twisted arm—which one?

Probe notes should be crossed out after they have been covered. The interviewer should be comfortable at the end of the interview, in stating that he or she needs to check notes to make sure everything was covered.

EXERCISE 5.5

Probing Exercises

Activities: The students will read a written interview and identify probes used by the interviewer.

Purpose: To identify effective and deficient probes and formulate effective probes

Activity 1: The students should read the event surrounding the interview and then the interview. Paying particular attention to the objectives of the interview, review the probes used by the interviewer. If the probe is good, the student should state why it is good. If the probe is poor, the student should state why and make corrections or write another probe.

Interview incident: At a local nightclub, a fight broke out with two groups of people punching at each other. At some point, a knife was drawn, and one of the fighters was stabbed fatally. Two bystanders were cut seriously enough to be hospitalized. The nightclub was packed with many people on the floor dancing to loud music. The police arrived at the same time as the ambulances. After taking care of the medical emergencies, officers interviewed witnesses. The following is an interview of one of the witnesses who was not actually involved in the fight but was dancing nearby. The purpose of the interview is to obtain information about what the witness saw and heard as it relates to what instigated the fight, who was involved, and the degree of their involvement.

1a. Officer: I'm Officer Shelton with the Middletown Police Department. Quite a lot going on at the club tonight. Are you okay?

1b. Witness: You're not kidding. I just came in to relax, dance, and have a couple of beers. Yeah, I'm okay. What was the fighting about, anyway?

2a. Officer: That is what I am investigating. I'd like to get to the bottom of this so folks like you don't have to worry about coming into your local club and getting hurt. I need for you to tell me anything you saw or heard since you got to the club.

2b. Witness: I got here about 10 P.M. with a couple of friends. The bar was going full blast by then so it was hard to see anything that would lead to a fight.

3a. Officer: Let's start when you first arrived. When you came in, what did you first do?

3b. Witness: The bar is across the room from the entrance so I made my way across the room to the bar and bought a beer.

4a. Officer: Is this the club you usually go to? Did you see anybody you know?

4b. Witness: Yeah, I come here every Friday and sometimes Saturday nights.

5a. Officer: What happened after you bought a beer?

5b. Witness: I saw this really hot guy looking at me so I went over and introduced myself.

6a. Officer: What did you do then?

6b. Witness: We talked for a few minutes, and then he asked me to dance. So we were out on the dance floor. We danced that whole dance, and it was just over when we heard a crash, somebody cussing, somebody else screaming, and then about four or five people really started getting into it.

7a. Officer: Did you know any of the people getting into it?

7b. Witness: No! They weren't any of my friends.

8. Officer: Thank you for your information. I need your name and contact information in case we need to follow up with you.

Activity 2: Role play with a partner using different levels of topic control probes.

1. The instructor will give the students a list of topics from which to choose.
2. Each dyad of students will pick two topics on which they are comfortable being interviewed. Each interview should last about 5 minutes.
3. Each student should take a few minutes to prepare a beginning open-ended question to invite an unrestricted response. The students should have available pen and paper to make probe notes.
4. The student interviewer should begin with a broad, open-ended question to his or her partner. Use low topic-control probes (active silence and encouragement) to promote elaboration. After 2 or 3 minutes, the student should change to high topic-control probes (clarification and elaboration—immediate and retrospective).
5. The student interviewee should respond as spontaneously and candidly as possible.
6. After 5 minutes, the instructor should tell the students to reverse roles and complete the second interview.
7. After completing the second interview, the two students should discuss the two role plays covering some of the following information:
   a. Did both students correctly use low topic-control and high topic-control probes?
   b. Which types of probes were the more difficult?
   c. How did the respondent respond to each kind of probe?

The instructor may want to discuss the exercise with the entire class.

Suggested list of topics:

U.S. role in the Middle East
Gun control
The problem of AIDS
Global warming
The immigration issue
The perfect career
The perfect vacation

Conducting Helping Interviews

The court advocate looked at the young woman and smiled reassuringly, "I know you have been through a great deal. I don't want to ask you to repeat anything that has happened to you, but rather help you assess how you are emotionally. What resources do you need to get on with your life?"

Although the main purpose of all interviews is to gather information, helping interviews do not end with the collection of information, but rather begin the self-exploration process of the client. The interviewer ideally is beginning a long-term professional relationship with the interviewee, or client. Rapport building is extended to relationship building, and the information enhances the treatment of the client.

If the helper is going to beneficially influence the life of the client, he or she has to establish a basis for this influence. The client needs to see the interviewer as an expert who is helpful and trustworthy. The interviewer must be skilled in grounding the helping process in concrete feelings and behaviors and keeping the client in the present.

Skills to counsel troubled clients go beyond the purview of this basic interviewing book. Nevertheless, initial interviews of helping relationships are often conducted by criminal justice personnel, who then refer clients who need long-term counseling.

As with any interview, the interviewer will establish rapport. He or she must respond to the client in a way that shows the interviewer is listening and understands how the client feels and what he or she is saying. Because the purpose of the interview is to provide a forum for self-exploration, the importance of empathy takes on a new role. The interviewer must see the client's world from the client's frame of reference rather than the interviewer's frame. It is not enough to understand; that understanding must be communicated to the client. The client's feelings should be emphasized more in helping interviews.

"My mother wants me to move out at the end of the month. She says that the baby's crying wakes her up at night so she has trouble working during the day. I don't know what I am going to do. I don't want to live with Steve."

"You feel hopeless and overwhelmed and don't know how you and the baby are going to live after this month. It sounds as if you believe your only option is living with your baby's father. What is it about living with Steve that bothers you?"

"Steve still uses cocaine. Every time he gets a job, he only keeps it a couple of weeks because he stops showing up or begins stealing to feed his habit. I'm afraid that if I move in with him, he will use the money for drugs that I get for the baby and me. I can't believe anything Steve says. I can't trust him. I am not sure I even love him anymore."

"So you are scared and believe it would be dangerous for you and the baby to live with Steve. You realize that while the baby is Steve's responsibility as well as yours, you don't believe he will put the baby's needs first. It sounds like you have problems being assertive, making your needs and expectations known to Steve."

"I don't know what to say to him."

"We can role-play some different situations next time I come. We also offer classes on assertiveness training. Does that sound like something you might want to try?"

"Yeah, I never heard about stuff like that."

"Good, we'll try it next week, and I will bring you some information about the training. We can further explore your feelings about Steve."

The interviewer's feedback, through paraphrasing, should include content and feelings that are heard from the interviewee, moving gradually toward the exploration of critical topics and feelings. Because accuracy of the paraphrasing is important, after the interviewer responds, he or she should attend carefully to cues that either confirm or deny the accuracy of responses. Signs of stress or resistance should be assessed. If these signs emerge, the interviewer should judge where these signs of stress or resistance come from. Could it be a lack of accuracy on the part of the interviewer's paraphrasing?

Keeping the clients in the present is necessary. If clients attempt to talk about the past or future, they must be urged to identify their present feelings.

> "My parents always made me feel worthless. I could never do anything right"
> "Your parents didn't praise your efforts when you were young, just criticized you. So you feel that efforts you make now are not worthwhile. Let's explore some of those efforts. What is an example you can describe of a recent effort?"

If they use second or third person, e.g., "You can't get a fair deal" or "One never knows how much it hurts," they must be encouraged to use I statements. I statements and talking in the present help them to own their statements, e.g., "How do you feel at this moment?" "I'm fed up and disgusted with probation. I don't think it is helping me at all."

Helping the client discuss problems concretely will improve the client's ability to self-explore and lead to problem-related information and solution-oriented resources. The following are examples of broad, ambiguous statements and then concrete examples:

> Vague statement: Things were not so hot today.
> Concrete example: Several customers complained about my service today. Then my manager yelled at me because she said I took too long at break.

> Vague statement: People pick on me.
> Concrete example: Other kids make fun of me for being fat. They call me "Porky" and "Fatso."

> Vague statement: I mess everything up.
> Concrete example: Instead of staying home last night and doing homework, I went out. So today I failed a test and didn't have a report ready for history that was due today.

> Vague statement: My relationship with my father bothers me.
> Concrete example: My father phones me at least twice a week. I get so tired of talking to him about his problems, but I know he doesn't have anybody else to talk to so then I feel guilty for getting irritated.

One of the areas that criminal justice professionals focus on with their clients is unhealthy assumptions that often underlie unhealthy behavior. An example is working with an adolescent who assumes all adults are as disapproving and intolerant as his parents. This assumption contaminates his interaction with all adults: "You have distrusted adults so long that you're still wondering whether you can trust me. You're still not sure, and so you are hesitant to share your feelings with me."

Often, clients have unrealistic or unformulated goals. For example, a probationer wants to become a lawyer but does not have a high school degree. The professional helper supports the probationer to explore and learn more about himself or herself. The professional helper has to begin where the client is at the beginning of the relationship, but then must help the client get a new perspective on his or her life and behavior. Self-exploring includes an inventory of the client's resources as well as defining the barriers to making changes.

The classic questions are as follows:

- How does the way the client currently respond to the problem work for him or her? Does it get the client the results he or she wants?
- Does the client want a different result?
- What does the client need to change to get the different result?
- What are the realistic steps the client needs to take to achieve the different result?
- What are some of the resources the client has internally to help achieve the different result?
- What are some resources that the client has externally to help achieve the new result?
- What are some resources that the client might need to obtain to help achieve the new result?

Resistant Clients   Another common problem for criminal justice professionals are resistant clients. Many clients have been referred, more often ordered by the court, to see the interviewer. These reluctant folks start off with a bad attitude and are unwilling to participate or cooperate in the helping process. Their unwillingness manifests in holding back, disengaging, or undermining change efforts. The clients may make appointments and then just sit and look out the window.

In domestic-violence situations, there are complex dynamics between two disputants, who are intimately involved. Often the violent episodes are used subconsciously by one or both disputants to preserve their relationship. As one partner begins to grow, the other partner might attempt to undercut the counseling.

Resistant clients may directly state they do not want help. They will tell the interviewer they do not have any problems, and/or it is not their fault. More subtle behaviors may be forgetting meetings, coming late, changing the subject when the discussion focuses on their problems and issues, and talking about superficial issues. Other resistant clients may rationalize their behavior or display attitudes of helplessness. They might become victims.

To minimize resistance, the professional needs to be sure the purpose and goals of his or her interactions are clear, each of his or her roles are clear, and all contracts are thoroughly understood. The professional needs to listen and convey acceptance, invite questions, and discuss reservations.

From the beginning, self-determination of the client should be fostered. Whenever signs of resistance are observed, the professional makes a specific observation and explores it with the client, such as, "I notice you are frowning. Do you have a question or some concerns about this?"

The client should only be confronted when there is no change or progress over a period of time. The purpose of confrontation is to encourage the client to change the frame of reference from which the client sees the world and his or her problems.

The client is questioned about the lack of progress and invited to suggest possible reasons and means of making progress to stimulate problem solving: "We have been working on this for a while, and you have attempted a variety of strategies, but they don't seem to have worked so far. Why do you think this is so? What do you think you could do differently?"

Confrontation is only possible if the professional has established a positive, working relationship with the client. If used too soon in the helping relationship, the client may perceive it as judgmental and may feel threatened. Even with the best of working relationships, confrontation needs to be used carefully. The professional needs to be careful about the words and tone of the message.

When professionals perceive a lack of trust directed towards them personally, they should introspectively explore their words, actions, and behavior. What may have produced the mistrust? Is there a better way to approach this problem? Trust is harder to gain when clients are involuntary, or referred by the courts. Representing authority, the professional may be viewed by the client as an agent of the court system. Self-exploring may help in determining service needs as well as exploring feelings.

> "I know you are saying that you will help me, but people never mean what they say. They are just trying to trick you and get you to do what they say."
> "Really, has that happened to you before? Tell me about it."

It may be necessary to negotiate and bargain, clarifying what is and is not negotiable. The courts may mandate that the client attend weekly sessions, but the client may have some choices on what is accomplished during those sessions. Any time the client is allowed self-determination, it will provide growth. Can the client help establish the time, the day of the week, or the frequency?

> "I know you said that it wasn't your fault, that you are taking the blame for somebody else. But you are now under court order to come here an hour every week, and I'm required to send a report to the courts every month about your attendance at these meetings. You know if you don't come, you might be sent to jail. So what would you like to do during our time together?"

It is effective to request that the interviewee summarize what has been said in a courteous and diplomatic fashion. It is not to be used to see whether he or she understands, but rather what was understood, e.g., "Why don't you go back over what we discussed so I can be sure I've told you everything."

The summaries should occur in manageable units to ensure critical points were heard. These summaries guarantee that the important information is heard and are used to confirm contracts, clarify priorities, and substantiate action steps toward meeting the goals.

EXERCISE 5.6

Helping Interview Exercises

Activity: Students will practice empathic responses that include feelings and content.

Purpose: To help students identify feelings of clients and further client self-exploration

Activity 1: Decide if response is good (+) or poor (−). If poor, indicate reason.

Boy, 15 years old, to school counselor: Coach Jones has it in for me. We have never gotten along. I don't do anything worse than anybody else on the team, but when there's a screw up, I'm the first one he yells at. I wish he'd get off my back,

1. ( ) You ought to chill. Why get thrown off the team for something stupid?

   Reason:_____

2. ( ) You feel he's being unfair to you—and that's miserable.

   Reason:_____

3. ( ) You've been in trouble before. Are you really giving it to me straight?

   Reason:_____

4. ( ) We can straighten this whole thing out if we all just chill. I think we're all reasonable people. By the way, how's the family?

   Reason:_____

Officer to a supervisor: I feel like an errand boy around here rather than a law enforcement officer. I get all these service calls to old ladies' homes. They have nothing better to do than call the police with every little noise they hear. I want to be doing real police work.

1. ( ) Why don't you stop your complaining!

   Reason:_____

2. ( ) You are irritated because you believe that you are not being full utilized as you have been trained.

   Reason:_____

3. ( ) I feel you have it in for me. You attack me because you can't get at the system. So you make me feel lousy.

    Reason:_____

4. ( ) I saw your wife and kids at church the other day. They really look great.

    Reason:_____

Resident of a halfway house, 19 years old, to a caseworker: You always hang around with the older guys, playing pool and ping pong. I have to sit in the dining room alone. When you do talk to me, I get the feeling you think you're wasting your time.

1. ( ) You're just jealous of the older residents.

    Reason:_____

2. ( ) Come sit with me now, and we'll work everything out. We can have a nice chat.

    Reason:_____

3. ( ) I really don't like the older residents at all. I have to force myself to be with them.

    Reason:_____

4. ( ) You feel lonely, and you think I don't feel you are worthwhile. Do you think that the other caseworkers also don't spend enough time with you?

    Reason:_____

Activity 2: Student triads will practice empathic responses with a partner. Students will select personal topics they feel comfortable sharing with two other people in class. They will discuss this topic with a partner. Using the assessment scale below, an observer will assess the partner's skill in handling feelings, pacing, keeping the speaker immediate and concrete, and summarizing.

1. The partner may want to begin with an opening question, "You appear to be quiet and thoughtful today. Is there something on your mind?"
2. The student will discuss a personal topic such as
    a. considering a change in major
    b. wondering if he or she said the right thing to a friend
    c. wondering what the best approach is to ask a professor about an academic problem
    d. considering changing jobs

3. The partner should listen carefully and paraphrase thoughtfully, identifying the content of the issue, how the speaker is feeling about it, and using a tentative phrasing to allow for correction.
4. After 5 minutes, the instructor should tell the group to stop. The observer should ask the student how he or she felt about the partner's paraphrasing. The student is allowed to comment. The observer then provides feedback using the assessment scale.

Assessment Scale

Skill	1	2	3	4	5
Handling feelings expressed by interviewee	Ignored feelings completely or went off in an irrelevant direction	Some attempt, but inaccurate	Accepted feelings, but was neither good or poor at making meaningful reflections	Accepted feelings and generally was able to make meaningful reflections	Accepted feelings, was able to make meaningful reflections at all times
Pacing	Questions asked too quickly, interview rushed	Interview hurried, questions or responses not paced, few pauses	Average speed and pacing	Well-paced questions or responses at appropriate times	Excellent use of time with well-paced questions and responses
Paraphrasing	No use	Attempted, but not effective	Paraphrased content information effectively	Paraphrased feeling information effectively	Paraphrased both content and feeling information effectively
Concreteness and getting specifics	Vague, generalities, abstractness	Rather vague and nonspecific	Neither noticeably vague nor specific	Fairly specific and concrete, avoided generalities and abstractions	Very specific; concrete actions, feelings, thoughts
Immediacy	Allowed interviewee to stay in the past	Seemed to recognize that interviewee was in past, but didn't attempt to change	Attempted once to bring interviewee into present	Some success in bringing interviewee into present	Kept interviewee in present throughout interview

Conclusion

Conducting interviews requires skill and experience. This chapter includes some of the techniques and information to help an interviewer acquire the skill necessary for effective interviewing. Only practice will provide the experience.

Initially, it is important that interviewers analyze their interviews so they can learn from each one. Some of the self-assessment questions that should be asked are as follows:

- Did I establish rapport?
- Did I motivate the respondent to provide complete, relevant, and valid answers?
- Did I actively listen to the interviewee using attending behavior, reflecting his or her content and feelings behind the message?
- Did I use appropriate nonverbal behavior?
- Did I effectively use silence?
- How effective was my probing? What was the level of topic control?
- What will I do differently on my next interview?

Helping clients requires a deeper level of empathy than investigative interviews. The professional must first respond to the client's world to help the client explore himself or herself. As the client begins to shape a more objective self-picture, change can begin. The change must come from the client, but the professional supports and helps the client choose constructive behavioral goals. The professional is a source of support, motivation, and modeling for the client.

ALLEN'S WORLD

Dr. Lord touches on some crucial elements of interviewing. She is dealing with nuances, innuendo, empathy, and interpretation of body language and gestures. All of these are critical to a private investigator. I need to know what information is valid, or deceptive, misleading, or a lie, either by omission or verbalization.

My needs are the same as Dr. Lord's. Get the accurate information. When my client is the defendant charged with a serious crime, my need for information is critical. Someone's life or liberty often depends on how thorough I am at information gathering.

Some of my tricks include the following:

If the subject answers my question with a question, I call that a nondenial denial or a nonanswer answer. If someone says to me in response to a question, "Why would I know that?" I smell a rat. And, I dig further.

Never tell someone, "I know how you feel." Nobody can discern with certainty how another person feels about anything. That is why Dr. Lord says to phrase reflections of feelings with tentative words that allow feedback. People's feelings are created by their background, which, in most cases, is going to be different from your background. When the interview calls for empathy, tell the subject, "I know how I would feel if that happened to me."

I once told a woman, whose husband of many years had just died, how sorry I was. Rather than the response I expected, she exuberantly told me that her husband had gone to a better place—and she couldn't wait to join him.

I once told a mother whose son had just drowned that I was sorry for her loss. "I told him to stay away from the pond," she responded angrily. "I tried to get him to take swimming lessons. He refused. He deserved what happened to him."

Again, you can never know how someone else feels about an event. All you can do to empathize is tell them you know how you would feel if that event happened to you.

Always tell subjects, "No hearsay." Only relate what they saw or heard, not what someone else told them they saw or heard.

Subjects often want to be helpful when they cannot remember. By doing so, they create problems for themselves, the investigator, the attorney, and the client. Always tell a subject the best answer might be, "I don't remember." That is because whatever case I am working on, that case might not go to trial for years. If they cannot remember an incident, a fact, or a detail shortly after the event, how can they possibly remember it years later?

Make note of outright inconsistent statements. I recently interviewed a client charged with extortion. He had a 2-year affair, videotaped some of the trysts, and the woman ultimately filed a complaint of extortion when he told her she could have the tapes back for $100,000. During the interview, he told me, "I love my wife, but . . ." There was little doubt, and I told his attorney that I thought our client was lying.

In general, when people are giving a response, and they use the word *but*, I disregard everything that came before. So, when my client said, "I love my wife, but . . ." I did not believe anything he told me before that.

When probing, I often, intentionally, read back incorrect information. Is the subject paying attention? Can the subject remember what lie he told, or what he omitted? Reading something back that I know is false helps discern fact from fiction. It creates an emotional reaction that disarms the subject.

For example, early in an interview, a potential witness told me he left his house, naming the three friends who were with him. Later, when I began probing, I read back the names of two of the correct people but threw in the wrong name for the third person. The witness did not correct me. At that point, I knew he was lying.

If the witness is telling the truth, most often he or she will get angry, point a finger at me, and say angrily, "I didn't say that." Regardless, it disrupts their rhythm.

Learn a form of shorthand. It makes note taking easier. For example, B4 instead of *before,* 2 instead of *to* or *two,* U instead of *you,* and 4 in place of *for.*

Dr. Lord and I agree about the need for probing and when to probe.

During the interview, I take notes of follow-up questions. I let the client or witness talk freely, planning to follow up at the end of their discourse.

However, clients ramble at times. They get off track. Knowing when to interrupt and break the free flow is difficult. If the information coming out is irrelevant, I will ask a probing question to the get the interview back on track.

Interview Documentation

Upon completion of this chapter, students should be able to

1. Consider a variety of tools for documenting the interview
2. Facilitate the writing of a complete statement by the interviewee
3. Prevent inhibitors to complete statements by interviewees
4. Interpret areas that need to be probed or investigated further after reading a statement

Introduction

Along with the variety of tasks the interviewer must handle is a means to document the interview. No matter how accurate the interviewer's memory is, written or taped documentation must be utilized. Documentation ranges from brief note taking during the interview to the written statement of a suspect. How comprehensive the documentation is depends on the type of interview conducted. For example, if working for a defense attorney, interview notes and statements may be turned into a formal affidavit, and the interviewee will need to sign it.

This chapter begins with tips on taking notes during the interview to help the interviewer remember areas to probe before the completion of the interview and to include in the interview report after its completion. Even videotaping or audiotaping of interviews does not preclude the need for notes. Learning to take good classroom notes for any college course will help students learn how to take notes during interviews. Simultaneously listening to an instructor and writing key points compare favorably.

After a thorough investigation, the statements from alleged offenders are the final, but critical, step. An interviewer would not want to complete the suspect's interrogation and then be faced with a refusal to write a statement. Another damaging outcome would be a written statement from the perpetrator that did not include introductory information that denoted the statement was given freely and voluntarily. The statement is an integral step in investigative interviewing.

The importance of properly structured statements has been amplified in recent years with the introduction of investigative discourse. Besides the overview provided in this chapter, readers may study the scientific use of stance analysis in Chapter 13.

Recording and Note Taking

Ideally, all interviews would be videotaped so interviewers could focus entirely on the interviewee. Digital videotaping is regularly used for interrogations, thereby avoiding challenges of voluntariness by the suspect, accuracy of the interview, and completeness of confessions (Shearer, 2005). Videotaping interviews, much like in-car videos used in vehicle stops, protect officers from false accusations but also lead to better preparation of the interviews. The possible viewing by supervisors and the courts has led to more conservative interviewing.

Unfortunately, interviewers often are forced to record the interview by using written notes. Even with recorded interviews, the interviewer should take some notes as insurance against faulty equipment. Notes for probes also are essential. Yet the most careful note taking may distort the information given or be selective by only including those portions that subconsciously agree with the interviewer's attitude or preconceived thoughts (Fisher, 1995).

Note taking also is distracting. The interviewee will want to know what the interviewer is writing. The interviewee may begin to watch when the interviewer takes notes rather than concentrating deeply on what the interviewer is asking. The interviewer needs to be as inconspicuous as possible, using some form of abbreviation or shorthand. Visual focus needs to remain on the interviewee rather than on notes.

In helping interviews, distortions in recording the client's responses could seriously restrict the interviewer's understanding of the client's problems (Walsh, 1988). Often in helping interviews and counseling sessions, the interviewer makes notes from memory after the interview and after the client has left. This strategy is effective only when small portions of the conversation are relevant. The interviewer also must have a good memory and be astute enough to recognize what information is relevant and valid.

Taking at least minimal notes and expanding them after the interview is helpful. These notes are useful especially for probe notes. The interviewer can still pay attention to listening and observing.

If the interviewee speaks quickly and provides a great deal of relevant information, then there is a danger that much will be lost. Verbatim notes, writing down all significant points using the exact words of the interviewee, are difficult and can influence the interviewee's motivation to provide complete responses. The interviewer does not want to be in a position to ask the interviewee to slow down or repeat what he or she is saying. The interviewer must be able to concentrate on listening, observing, evaluating, and probing, so it is important to identify interviews considered sufficiently critical to bring in a transcriber who can capture the entire interview as if it were recorded, including audible nonverbal cues, pacing, hesitations, interruptions, and false starts.

Probe Notes

"I've been working real hard to find a good job." (The interviewer scribbles quickly on a note pad, "good job-define.") "I don't want some hamburger-flipping joint. I know I can do better than that. I've been checking for open positions that need my skills." (What sources has he checked? Describe skills).

Probe notes are taken to remind interviewers of specific points in interviews that need elaboration and clarification.

As discussed in Chapter 5, the interviewer does not want to interrupt the interviewee and risk disrupting his or her train of thought. Probe notes allow the interviewer to remember areas to return to at an appropriate pause in the interviewee's thoughts.

Good probe notes spring directly from the process of evaluating the relevance, completeness, and validity of the responses to direct questions. They should just be a word or phrase that does not take away the interviewer's attention from the ongoing interview.

It is useful to make the note using exact words of interviewees so the probes remain understandable. In the above interview, the interviewee states that it is hard to find a good job. The interviewer will ask a follow-up probe about what the interviewee considers a good job, using the same words. If the interviewer uses any other word besides *good*, then the interviewer already is beginning to define the meaning of good job rather than getting that information from the interviewee.

As the interviewer asks a follow-up probing question, then the probe note should be crossed out. At the end of the interview, it is appropriate to tell the interviewee, "Please give me just a minute to look over my notes. I want to be sure that I am clear about everything you have told me." The interviewer then glances through all of his or her notes to be sure that everything was covered sufficiently.

EXERCISE 6.1

Practicing Probe Notes

Activity: In triads, students practice taking and using probe notes in short class interviews.

Purpose: To help students practice improvising probes with the use of probe notes.

Divide the students into groups of three. Each student should be given the opportunity to be the interviewer. The interviewees discuss who they plan to interview for their final class project. (The instructor can create any number of different topics the students might briefly discuss.) The interviewer does not have to think about questions that need to be asked nor actual probes, but just take notes on what needs to be probed further for clarification or elaboration. The third person, as observer, does the same. After the student who is the interviewee is finished, the interviewer and observer compare probe notes.

What notes were similar? What notes are different or additional? Did they remember to include actual words of the interviewee?

Variation of exercise: Divide students into pairs. Students take notes of what each other say during an interview of 3 to 4 minutes. The interviewees can talk about anything, or the instructor can give them a topic. The interviewer is to practice active listening and taking notes simultaneously. After each student in a pair has the opportunity to be an interviewer, the instructor has the students discuss their experience.

Could they listen actively?
Could they assess for relevancy, completeness, and validity?
Could they observe the interviewee's nonverbal cues?
Were they able to reflect content and feelings of the interviewee?
Were they able to write accurate notes? Did they use shorthand?

The Statement

"Now, Ms. Reimer, I have taken a great deal of your time, and I appreciate your cooperation. You have provided a lot of information. I have one last thing I need for you to do. I need for you to write down what you have told me. We have covered a great deal, and I need for you to write down what happened in as much detail as possible. Please select some point in time before the incident and write down all of the events, who all were involved, and your feelings surrounding the incident. Take your time and let me know if you need a break."

A statement is a written document or recording obtained from a victim, witness, or suspect that covers all the information the interviewee has given during the interview. The statement has several purposes:

1. It locks the interviewee into the facts detailed during the interview.
2. It prevents the interviewee from changing stories.
3. It serves to refresh the interviewee's memory if at a later date, he or she must recount the story.
4. It acts as evidence if the interviewee dies prior to a court hearing, is unavailable, or becomes mentally incompetent (Zulawski & Wicklander, 1993).

The importance of the written or recorded word is clear to most people. Interviewees, willing to describe incidents, suddenly become silent when asked to recount the information as a written statement. Statements that can be proven false have almost as powerful an effect as a confession of guilt on juries. So even in cases that the interviewer is fairly sure the interviewee is lying, he or she should be encouraged to write a statement detailing the lie regardless of how insignificant it may seem at the time. Encouraging as much detail as possible helps the investigator–interviewer find evidence to support or disprove the alibi. Obtaining statements from multiple suspects or witnesses is helpful, especially when separating fact from fiction.

Civil or criminal cases often take 2 or more years to proceed through the courts system. Clarity of detail begins to fade over time. A statement written by victims and witnesses should contain as much detail as possible because 2 or 3 years later at trial, they can use their statements to help them recall details. Under normal circumstances, the statement never stands alone at trial. However, if the statement writer has died or becomes mentally incompetent, the statement might be allowed to be entered as evidence.

The most common type of a statement is a narrative, which is a handwritten account by the interviewee using first person, *I*, to describe activities in the incident, as well as the people involved. It is used to substantiate involvement or what was observed and often contains elements of the crime, personal feelings, and the interviewee's state of mind at the time of the crime or incident.

Another type of statement, the question-answer narrative, is rarely used alone, but rather as a supplement to the narrative statement first completed by the interviewee. Questions asked by the interviewer clarify certain points. The interviewer's questions and the interviewee's responses are added to the end of the original statement.

Formal Statements

"This is Detective Caldwell, Greensburg Police Department. This is a taped statement from a Denise Amy Broughton. The time is 1130 hours on February 10, 2009. Ms. Broughton resides at 2702 Appleton Lane, Greensburg, Maryland. She is the owner of the Cat's Meow Club on Briar Creek Avenue. The incident occurred at the Cat's Meow on February 9, 2009. Ms. Broughton, please state your name for the tape."

If the incident is serious or potentially costly to an agency, the statement should be videotaped, audiotaped, and/or typed by a court reporter or authorized stenographer because most witnesses, victims, or suspects will shorten a written statement by omitting valuable information when the writing becomes emotionally difficult. Statements usually come at the end of several hours of interviewing. At a minimum, the interviewee is exhausted and unwilling to write out page after page.

The interviewer should have invested some time in the interview and with the interviewee before beginning the formal statement recording. Among the many reasons that the time investment is critical is the importance of knowing already what the interviewee will be saying in the formal statement.

It is especially important that the victim has had time to become calm and has already recounted his or her story, providing all the details that the interviewer needs to have a complete, relevant, and valid interview. The tape is likely to be played if there is a trial. The jury will be able to see and hear the state of mind of the victim at the time of the statement. They also will be able to assess emotional states and voluntariness of any admission.

When to begin the recording is a difficult decision; even which interviews to videotape becomes a resource issue for agencies. Legally, there can be a question of the timing to videotape if the entire interview is not taped. Interviews can range from 2 to 4 hours. If they are not completely taped, questions emerge about what occurred before the taping began. Also if the recording is turned off or tapes are changed, careful explanations should be included, "Tape one is completed. Tape two is beginning. No questions or comments were made in transition."

At least one large police agency mandates videorecorded and taped interviews for suspects involved in murders, sex offenses, armed robbery, and kidnapping. The video recording must be activated prior to the interviewer entering the room to conduct the interview and must remain on until the suspect is permanently removed from the room (Charlotte-Mecklenburg Police Department, 2009).

If any of the witnesses or suspects are reluctant to be recorded, the interviewer should remind them that the sound of their words reflects the truthfulness of their descriptions or their side of the story. Instead of using the word *statement*, the interviewer should use *explanation*, thereby reducing the formality and consequences associated with words like *confession* or *statement*.

When interviewing reluctant suspects, the interviewer should sell them on the chance to tell their side of the story—to explain mitigating circumstances. Witnesses can be encouraged by suggesting they will less likely be inconvenienced with additional interviews after providing a statement or written explanation.

However the statement will be conveyed, the interviewer should provide direction, controlling the format without actually dictating the words and never leaving interviewees alone while they are relaying their explanation. The statement should be taken immediately at the conclusion of the interview, before either interviewee or interviewer leaves the room. If the interviewee is reluctant to write or explain his or her story on a recording, it is best to get a witness to hear oral statements.

Statement Format   During formal statements, the interviewer must not tell the interviewee what to write, but can guide the formulation of the statement. Areas of importance include the introduction, substantiation of the interviewee's role, and information related to the offense or incident.

The introduction should include the date and time of the interview and biographical information such as the full name of the interviewee, home address, job title, and place of employment. This information is not considered threatening. If it is a suspect's statement, then it should include a verification of voluntariness of the statement with the Miranda waiver included.

As with the interview itself, the interviewer suggests that the interviewee begin with details proceeding the incident or crime. The interviewer must ensure information related to the elements necessary to prove the crime is included. For example, if money was stolen, the statement should include the amount stolen, method of theft, what the money was used for, and the location of any evidence to the crime or remaining money, if known by the interviewee.

Especially with suspects and reluctant witnesses, the interviewer will need to encourage the interviewee to detail involvement. It is useful to have the details of involvement near the beginning of the statement, so if the interviewee refuses to finish the statement, it already contains involvement, intent, and/or the elements of the offenses.

Narrative statements in which the interviewee provides details are preferred. If the interviewee completes the statement and there are gaps, then the interviewer should include a supplemental question-and-answer statement. The interviewer should write the question exactly as it was given, accompanied by the exact response of the interviewee. Much like the use of probes, the interviewer may need to ask questions that clarify terms used by the interviewee or that expand portions of the statement.

Once completed, the interviewer asks the interviewee to review the statement and to assure that everything written is true. If any corrections or clarifications are made, the interviewee is asked to initial each correction. If the statement is several pages, it is a good idea to have the interviewee initial each page. Each page should be numbered. The interviewer can reassure the interviewee: "I want you to initial each correction and each page so you are confident that your statement can't be altered."

Signature and Correcting Errors

All statements should be signed and dated. Formal statements also should be signed and dated by a witness, but only to confirm that the subject acknowledged that he or she wrote the statement (Inbau, Reid, & Buckely, 2004).

Sometimes interviewees will hesitate before placing their signatures on a document. The interviewer should anticipate the hesitation and ask: "Everything you put down is true, right?"

More than likely the interviewee will acknowledge in the affirmative: "Good. In writing your signature, you are attesting that you told the truth."

Challenges to Statements

It is not unusual for interviewees to willingly talk to the interviewer but then balk when requested to write down what they have revealed. There is something powerful and scary about seeing their words written down.

Realistically, they know that it is difficult to deny their words if they are on paper. Even when the statement is completed and signed, the interviewee may have second thoughts later and may attempt to deny the statement. The interviewer needs to be prepared to respond to the initial fears and hesitations, as well as the other possible challenges to the legitimacy of the statement.

If the refusal comes at the beginning of the request to make a written statement, it is possible that the interviewee is embarrassed about lack of writing skills or may even say he or she cannot read or write. Use of audio-taping or videotaping is particularly useful in these cases. The statement can be taped, and an authorized stenographer can transcribe it later. It is vital that the tapes remain protected—chain of custody—and a part of the case file so any question of possible editing of the tapes can be answered un-equivocally: "No, these tapes have not been doctored."

It always is a good idea to have an uninvolved witness during the statement production, whether the statement is to be written or recorded. The witness becomes even more important if the interviewer or an authorized stenogra-pher is writing the original statement for an interviewee who cannot read or write. The witness needs to be able to testify that the interviewee was treated properly, the statement was voluntary, and what was written was exactly what the interviewee said. Of course, the witness needs to sign the end of the state-ment acknowledging that the statement was voluntary and details exactly what was said by the interviewee.

If the witness is brought in after the initial interview, then the interviewer should summarize what the interviewee has said, asking the interviewee if the summary is accurate. The interviewer also will ask the interviewee to describe briefly the treatment he or she was given and the voluntariness of the statement.

The interviewer may be able to anticipate the unwillingness to write the statement and bring in a witness at the beginning of the interview, thereby preventing the possibility of a complete refusal that may come with a delay in time between the interview and the request for a statement.

If the interviewee is orally providing the statement for a stenographer, then upon completion of the statement, the interviewer should review the complete statement with the interviewee. Any corrections should be ini-tialed by the interviewee, and then the statement signed by the interviewee. If possible, the person writing the statement may want to make a few unim-portant errors so the interviewee's initials will be on the statement one or more times to show that it was carefully reviewed. It also prevents allegations that the interviewee just signed what was put in front of him or her.

At the beginning of the statement, there should be a sentence that records the individual who is writing the statement, for whom it is being written, and the date. Then, the next sentence should state that the statement is being given voluntarily. As will be further expanded in Chapter 10, if interviewees are suspects who are in custody, their Constitutional rights must be waived knowingly and voluntarily before they can be questioned. This mandate is so critical that when the case goes to court, the prosecutor bears the burden of proving beyond a reasonable doubt that the defendant's waiver is valid (Klot-ter, Walker, & Hemmens, 2005). If the suspect's acknowledgement of his or her rights and waiver is in writing, the prosecutor's job is much easier.

It is critical that if the statement is recorded by a party other than the inter-viewee, it is written verbatim, exactly as dictated by the interviewee. Grammar, syntax errors, and fragmentary comments should not be corrected.

Interviewer's Report

The interviewer needs to complete a detailed report of the statement process. This report is especially critical if the statement contains admissions and/or a confession. Any indication of involuntariness or illegal coercion by the interviewer may keep the statement from being introduced into court.

Although less emphasis is placed on reports completed by interviewers after other kinds of investigative interviews or helping interviews, it is still important for them to be documented. Helping professionals need to ensure a chronology of clients' progress and issues. Frequency of visits, cancelled visits, and monitoring activities such as drug tests must be carefully documented. Written records of all clients' meetings are critical.

The purpose of this book is to instruct students in the skills of effective interviewing, so its authors do not attempt to cover the complex and legal process of interrogations (Chapter 10 will provide an overview of the interrogation literature.). All interviews, not just interrogations, must be documented thoroughly. Statements especially should be safeguarded because they are symbols of protected individuals' rights in the United States.

EXERCISE 6.2

Practicing Taking Statements

Activity: Students role play introducing each other to the need to write a statement.

Purpose: To give students nonthreatening practice developing their confidence in requesting statements.

Students are divided into dyads and instructed for one to begin as the interviewer and the other, the interviewee.

In the first exercise, the interviewer has completed interviewing a witness of a store robbery. The interviewer now must persuade the witness to write a statement. The witness is tired and impatient. The interviewer must consider techniques to receive a complete statement. After the witness is convinced, the student who played the witness provides feedback on what seemed persuasive and what did not help.

In the second exercise, the interviewee is a disputant who got into a fight with another individual at a nightclub. The disputant wants the police to arrest the other party, but he is reluctant to take out an assault warrant. The interviewee initiated the complaint and told the interviewer his or her version of what happened. The interviewee states that he or she is not a very good writer. As with the first exercise, the interviewer must consider techniques to cajole a complete statement. After the witness is convinced, the student who played the witness provides feedback on what seemed persuasive and what did not help.

Investigative Discourse Analysis

To this point in Chapter 6, statements have been described as the written conclusions of interviews in which victims, witnesses, and suspects have provided accurate and useful information. Even in these statements in which interviewers believe interviewees will write complete and accurate information, there is a possibility that they will omit or "bend" the truth.

In addition, there will be a number of individuals who will have developed an alibi and openly provided the alibi during an interview and in a statement. These individuals can be victims or witnesses, as well as suspects for reasons already discussed.

Besides documentation for an incident or client file, statements can be used to further move an investigation along or to expand the information base on a client. Based on discursive psychology, *investigative discourse analysis* (IDA) is evolving to help interviewers understand what the interviewee really is saying in his or her statement.

Discursive psychologists believe thoughts can be interpreted through individuals' conversations and text. Cognition and reality are reformulated into discursive composition. Teaming up with linguists, discursive psychologists now use narrative analysis to dissect individuals' statements into structural categories.

Discursive psychology with narrative analysis has been used extensively to study sex offenders' statements (MacMartin & LeBaron, 2007; Riessman, 1993). Boyd Davis and Peyton Mason (2008) have further developed scientifically the analysis process into the pioneering field they call *stance analysis* (see Chapter 13).

In its broadest sense, IDA is the close and systematic study of the basic components of written communication to examine its validity and completeness.

All languages are constructed based on rules accepted by a set of people. These rules are internalized as people learn to speak. Although the rules may vary among languages, they all have basic components that detail an action or state of being; the subject of that action or state of being, usually an object of the action; and a variety of descriptors or modifiers. In other words, all languages contain the parts of speech such as noun, pronoun, verb, and adverb.

In formal education, students learn the grammar of their language and overtly learn how to properly formulate their written language; however, even illiterate people have internalized most of the rules behind the grammar. They may not be able to identify a noun, pronoun, verb, adjective, or adverb, but they know what sounds correct. It is true, as students might argue, that people often talk and write making grammatical mistakes, but the sentence structure they use will still be accurate for their language. Using *don't* rather than *doesn't*, or *ain't* rather than *isn't* does not change the general meaning or structure of a statement.

Early work in this area began in the 1950s by German psychologist Udo Undeutsch and Swedish psychologist Arne Trankell, who both developed *statement reality analysis* (Tully, 1999). Their work was expanded by Steller and Kohnken in the 1980s, who labeled it *criterion-based content analysis.*

Don Rabon (1994), in his book, *Investigative Discourse Analysis*, provides an extensive discussion of the analysis process. As Rabon states, to understand IDA, the interviewer does need to understand the basic parts of speech, grammatical rules, and the importance of word usage. Using IDA, the interviewer examines the statement's structure, or form, and then the meaning of the words used.

A note to caution readers: IDA is a tool much like a probe to help interviewers obtain complete, valid information. After thoroughly understanding this tool, interviewers can use their analytical results to further probe. An IDA cannot be considered an absolute judgment on whether the interviewee's intent was to fabricate information or to mislead the interviewer. Also, there are several steps that should be taken to formulate the analysis. Just as one piece of evidence does not lead to a conviction, one step of IDA cannot be interpreted as deception.

Analysis of the Statement's Structure

Statements have three parts: information about what happened before the event (prologue), what happened during the event (central), and what happened after the event (epilogue).

The first examination in an IDA is of the balance of these three parts. In general, the more equal the three parts are in length, the more truthful the statement is likely to be. For instance, if the interviewee omits information from one of the parts, then it will be shorter than the other two parts. On the other hand, if the interviewee adds information that actually did not occur, then that component of the statement will be disproportionately longer. When these imbalances occur, the IDA analyst labels it *deceptive on its form.*

To assess the balance, the interviewer must first determine the central issue—the factor that gives meaning to all of the rest of the narrative. Without an identified central issue, either nothing else in the statement would have occurred or would have mattered. If the interviewer has requested the statement about a specific incident, then the incident will likely be the central issue. The three parts of the statement then revolve around the central issue. What occurred before the central issue is the prologue, followed by the description of the central issue, and then what occurred after the central issue, is the epilogue.

Consider Senator Edward M. "Ted" Kennedy's now infamous statement about the 1969 accident at Chappaquiddick. On June 6, 1968, Robert F. Kennedy was assassinated. About 13 months later, on July 18, 1969, fate once again intervened in the tragedy-struck Kennedy family. Senator Kennedy left a party with Mary Jo Kopechne. His car plunged off Dyke Bridge (depending on the source used, *Dyke* is often spelled *Dike*) on Chappaquiddick Island into Poucha Pond. Kennedy survived. Mary Jo drowned.

Kennedy did not report the accident to the police until 9:45 the next morning. Senator Kennedy was asked to write a statement. The statement provides an example of a statement that is deceptive on its form (Damore, 1988).

The central issue or event is the car going off the bridge and sinking, which Senator Kennedy describes in lines 8 through 12:

01 On July 18th, 1969, at approximately 11:15 p.m. in
02 Chappaquiddick, Martha's Vineyard, Mass, I was
03 driving my car on Main St. on my way to get the ferry
04 back to Edgartown. I was unfamiliar with the road and turned
05 right onto Dyke Rd, instead of bearing hard left on Main
06 Street. After proceeding for approximately one-half mile on
07 Dyke Rd, I descended a hill and came upon a narrow bridge.
08 The car went off the side of the bridge. There was one
09 passenger with me, one Miss Mary, a former
10 secretary of my brother, Sen. Robert Kennedy. The car turned
11 over and sank into the water and landed with the roof resting
12 on the bottom. I attempted to open the door and the window
13 of the car but have no recollection of how I got out of the
14 car. I came to the surface and then repeatedly dove down to
15 the car in an attempt to see if the passenger was still in
16 the car. I was unsuccessful in the attempt. I was exhausted
17 and in a state of shock.
18 I recall walking back to where my friends were eating.
19 There was a car parked in front of the cottage and I climbed
20 into the backseat. I then asked for someone to bring me back
21 to Edgartown. I remember walking around for a period
22 then going back to my hotel room. When I fully realized
23 what had happened this morning, I immediately contacted the police
      (Damore, 1988, p. 22).

Lines 1 through 7 describe what Senator Kennedy was doing before the car went off the bridge. Part of line 12 through line 23 describes what he did after the car sank. Given the education of its author, the entire statement is brief and should leave the interviewer wondering what has been omitted.

Once the statement is dissected into its three parts, the interviewer can believe with confidence that the primary omissions occur during the description of the incident itself (18% of the narrative) and to some degree during the prologue (32% of the statement). As a tool, this first conclusion tells the interviewer that further questions need to be developed to expand what actually occurred that led to the car going off the bridge.

Mean Length of Utterance

Just as there is balance in the form, there is usually balance in the length of sentences. Although individuals vary in how long their sentences are, they themselves usually do not vary in length. Some individuals write in lengthy sentences, while others write in short sentences. At the point that the length of sentences changes, the statement is likely to be deceptive. For instance, the readers may have noticed the different styles of the two authors of this book.

To determine the mean length of utterance (MLU), the total number of words is divided by the number of sentences. For example, Senator Kennedy's statement uses 244 words. There are 15 sentences. When the 244 words are divided by the 15 sentences, the average sentence length is 16 words.

Using this approach, sentences that are particularly brief should be probed further. What is likely to be omitted in the nine-word sentence, "The car went off the side of the bridge," or the even shorter sentence, "I was unsuccessful in the attempt"?

The interviewer should be sensitive also to sentences that are abnormally long. Senator Kennedy's statement has a 26-word sentence that precedes the 6-word sentence. It is possible that the longer sentence contains fabricated information and should be probed further.

Semantic Analysis   Moving from the overall statement structure to the meaning of the words used, the interviewer adds another layer of analysis. Words that reveal the interviewee's lack of conviction or allusion of an action, as well as general statements, depersonalization, and use of present verb tense when describing an event that occurred in the past all indicate areas of the statement that need to be probed further.

Words that suggest the interviewee may lack conviction have been further refined in Davis and Mason's (2008) work on stance analysis. Strong statements such as "I told the truth" are stated with conviction. While it is possible for a few people to lie with passion and fervor that exude sincerity, most people will include modifying or equivocating words. Words such as *kind of, sort of, I don't know,* and *not really* soften the strength or conviction of the speaker. Consider the following:

> "These two guys approached me when I was putting gas in my car. I didn't really notice too much about them, but they sort of scared me. I kind of watched them because they seemed to be staring at me. I finished pumping gas and closed my gas tank. I reached inside my car to get my credit card. I was going to use it at the pump. As I reached inside, I think one of them grabbed my arm while the other one grabbed my purse. I don't know for sure. It happened so fast."

In the above statement, the reader can compare sentences in which the interviewee appears to make strong statements with those in which the sentences are softened with the equivocating words such as *really, sort of, kind of, I think,* and *I don't know.*

The above statement also includes a statement in which the interviewee alludes to an action without saying that she actually performed it. These statements often accompany the equivocating statements. The statement, "I was going to use it at the pump," is an allusion of an action. Arguably the interviewee included this statement to indicate the actions that she would have carried out if her purse had not been stolen. Because it was unimportant whether she was going to pay at the pump or inside the station, it provides a stronger case for allusion. The interviewee's lack of conviction needs to be probed further. If the interviewee uses strong statements throughout the rest of her statement, then there is further support that the interviewee has changed or omitted details. The interviewer should ask for more elaboration on the appearance of the attackers and the actions of the interviewee.

Senator Kennedy's statement provides another example of allusion of action. He states, "I then asked for someone to bring me back to Edgartown." He said he asked but does not claim to have actually gotten a ride back to Edgartown. In a later report, Senator Kennedy states that he swam back.

General comments in the middle of more specific sentences also are suspect of omission of information. Vague statements such as "talked for awhile," "messed around," "did paperwork," or "finished my job" often are used to gloss over what actually occurred. The interviewee may be editing the statement because he or she considers that information unimportant or wants to avoid the specific information. With either intention, the interviewer must probe.

> "Last Tuesday, I left work at 5:30 and drove home. I messed around the house for awhile until Susan showed up. We cooked supper together just making a green salad and putting a pizza in the oven. We ate and stuff and then left the house around 7:30 to go to the movies."

What does "messed around the house" or "ate and stuff" indicate?

Returning to Senator Kennedy's statement, depersonalization occurs both about himself and his passenger. At the point of the actual incident, the use of *I* vanishes. He had been driving his car. He was unfamiliar with the road. He descended the hill, but suddenly it was the car that went off the side of the bridge. The avoidance of self at the point of responsibility or possible involvement is a red flag to be followed. Tracking when *I* disappears and reappears enhances the interviewer's ability to interpret the interviewee's attempt to cover personal involvement.

Other examples of depersonalization are the changing of people to objects. Returning again to Senator Kennedy's statement, he originally introduces his passenger by providing Mary Jo's name and her association to him. After the car goes off the bridge, Mary Jo becomes *the passenger*. The use of a different term of reference should cause the interviewer to ask what caused the change in the relationship. Consider the following example:

> "I was worried when Jane didn't come home from work. Jane is prompt, and there is little traffic between her work and home. When she was an hour late, I went looking for her. I found the car about a mile from the house. About ten feet from the car, I found the body."

The interviewee personally identifies Jane until the point of finding the car. The car is not called *her car*, and Jane becomes *the body*. As noted several times, care has to be taken when interpreting for deception. A grieving person may need to remove himself or herself from the horror of the situation and deny the found body is his loved one. The interviewer still needs to be alert to the possibility of an individual attempting to distance his or her involvement from the criminal act.

Use of present tense when describing an event that occurred in the past should be considered suspect, especially if the interviewee shifts into present tense at particular places within the statement.

The reader may want to review the section on neurolinguistics. When people make up certain parts of their story, they will remember the actual facts and then decide what they construct. Watching individuals' eyes when they remember an event is compared to when they remember an event and then shift to constructing a modification of the truth. Similar action occurs during the telling or writing of the event with the use of verb tense. The interviewee will tell the past occurrence using past tense, but may carelessly switch to present tense as they are constructing a new truth.

> "On Wednesday, I wrote out all of the payroll checks as I usually do. After confirming that I had recorded all of them in the accounting book, I went to all of the employees and give them out. Susan isn't at her desk, so I put her check in her center drawer. I was going to tell her that I put it there when I returned to my office."

The interviewee remains in the past tense until the point in which the checks are given out. In the last sentence, the interviewee returns to past tense, but as readers will notice, the last sentence also alludes to an action (going to) without confirming it occurred. The interviewer should note the switch to present tense at the point that the checks are distributed, with special attention to Susan's check.

The emphasis of this book is on preparing for and conducting effective interviews that are relevant, valid, and complete. IDA is a backup tool that interviewers use if they receive, as part of their preparation, a preliminary statement from interviewees.

The analysis helps them prepare for the interview and to probe what facts may have been omitted or which details may have been fabricated. Even with well-prepared interviews, interviewers should closely review the statement that the interviewee writes. Using IDA techniques may reveal an area the interviewer missed.

EXERCISE 6.3

Using Investigative Discourse Analysis

Activity: The students complete an IDA on an actual statement as homework and then discuss it in class.

Purpose: To try some of the techniques discussed. Using an actual statement of a convicted offender helps the students understand some of the described techniques. It also should help them build caution in its use as a tool and not as an absolute.

Using IDA, consider such steps as (1) deceptive on its form, (2) MLU, and (3) different types of semantic analysis.

Captain Jeffrey MacDonald's Transcript

Early morning of February 17, 1970, the dispatchers at Fort Bragg in North Carolina, received an emergency call from Jeffrey R. MacDonald, a Green Beret physician. The military police and medics went to his house. They found MacDonald in his bedroom, alive, but wounded. Next to him, they found the body of his murdered wife, Colette, 26, who was pregnant with a male fetus.

Medics found his two daughters in their bedroom. They had been murdered.

MacDonald's most serious wound was a small, sharp incision that caused a lung to partially collapse. MacDonald was transferred to a hospital and released by the next week.

MacDonald gave a long, detailed version of what happened. There were intruders and a struggle. He was knocked unconscious.

The physical evidence did not corroborate MacDonald's story. There was little sign of a struggle in the living room where he said he had to fight off the three male intruders. Fibers from his torn pajama top were found under the body of his wife and in his two daughters' bedrooms, not in the living room. (Captain MacDonald's statement will appear again in Chapter 10.)

The following is the statement given to the detectives by Jeffrey MacDonald on April 6, 1970. He was formally charged with the deaths of his family on May 1, 1970 (*United States of America v. Jeffrey R. MacDonald*, 1979).

> Let's see. Monday night my wife went to bed, and I was reading. And I went to bed about somewhere around two o'clock. I really don't know; I was reading on the couch, and my little girl Kristy had gone into bed with my wife. And I went in to go to bed, and the bed was wet. She had wet the bed on my side, so I brought her in her own room. And I don't remember if I changed her or not; gave her a bottle and went out to the couch 'cause my bed was wet. And I went to sleep on the couch.
>
> And then the next thing I know, I heard some screaming, at least my wife; but I thought I heard Kimmie, my older daughter, screaming also. And I sat up. The kitchen light was on, and I saw some people at the foot of the bed. So, I don't know if I really said anything or I was getting ready to say something. This happened real fast. You know, when you talk about it, it sounds like it took forever; but it didn't take forever. And so, I sat up; and at first I thought it was—I just could see three people, and I don't know if I—heard the girl first—or I think I saw her first. I think two of the men separated sort of at the end of my couch, and I keep—all I saw was some people really. And this guy started walking down between the coffee table and the couch, and he raised something over his head and just sort of then—sort of all together—I just got a glance of this girl with kind of a

light on her face. I don't know if it was a flashlight or a
candle, but it looked to me like she was holding something. And I
just remember that my instinctive thought was that "she's holding
a candle. What the hell is she holding a candle for?" But she
said, before I was hit the first time, "Kill the pigs. Acid's
groovy." Now, that's all—that's all I think I heard before I,
I was hit the first time, and the guy hit me in the head. So I
was knocked back on the couch, and then I started struggling to
get up, and I could hear it all then—now I could—maybe it's
really, you know—I don't know if I was repeating to myself what
she just said or if I kept hearing it, but I kept—I heard, you
know, "Acid is groovy. Kill the pigs." And I started to struggle
up; and I noticed three men now; and I think the girl was kind of
behind them, either on the stairs or at the foot of the couch
behind them. And the guy on my left was a colored man, and he hit
me again; but at the same time, know, I was kind of
struggling. And these two men, I thought, were punching me at the
time. Then I—I remember thinking to myself that—see, I work
out with the boxing gloves sometimes. I was then—and I kept—
"Geez, that guy throws a hell of a punch," because he punched me
in the chest, and I got this terrific pain in my chest. And so,
I was struggling, and I got hit on the shoulder or the side of the
head again, and so I turned and I—grabbed this guy's whatever it
was. I thought it was a baseball bat at the time. And I had—
I was holding it. I was kind of working up it to hold onto it.
Meanwhile, both these guys were kind of hitting me, and all this
time I was hearing screams. That's what I can't figure out, so—
let's see, I was holding—so, I saw the—and all I got a glimpse
was, was some stripes. I told you, I think, they were E6 stripes.
There was one bottom rocker and it was an army jacket, and that
man was a colored man, and the two men, other men, were white.
And I didn't really notice too much about them. And so I kind of
struggled, and I was kind of off balance, 'cause I was still half
way on the couch and half off, and I was holding onto this. And
I kept getting this pain, either it—you know, like sort of in
my stomach, and he kept hitting me in the chest. And so, I let
go of the club; and I was grappling with him and I was holding his
hand in my hand. And I saw, you know, a blade. I didn't know
what it was; I just saw something that looked like a blade at the
time. And so, then I concentrated on him. We were kind of
struggling in the hallway right there at the end of the couch;
and then really the next distinctive thing, I thought that—
I noticed that—I saw some legs, you know, that—not covered—
Like I'd seen the top of some boots. And I thought that I saw
knees as I was falling. But it wasn't what was in the papers that I
I saw white boots. I never saw white, muddy boots. I saw—
Saw some knees on the top of boots, and I told, I think, the
investigators, I thought they were brown, as a matter of fact.

And the next thing I remember, though, was laying on the hallway—
at the end of the hallway floor, and I was freezing cold and it
was very quiet. And my teeth were chattering, and I went down and
—to the bedroom. And I had this—I was dizzy, you know. I
wasn't really—real alert; and I—my wife was lying on the—
the floor next to the bed. And there were—there was a knife in
her upper chest. So, I took that out; and I tried to give her
artificial respiration but the air was coming out of her chest.
So, I went and checked the kids; and—just a minute—and they
were—had a lot of—there was a lot of blood around. So, I
went back into the bedroom; and I—this time I was finding it
real hard to breathe, and I was dizzy. So I picked up the phone
and I told this asshole operator that it was—my name was
Captain MacDonald and I was at 544 Castle Drive and I needed the
M.P's and a doctor and an ambulance. And she said, "Is this on
post or off post?"—something like that. And I started yelling
at her. I said—finally, I told her it was on post, and she
said, "Well, you'll have to call the M.P's." So I dropped the
phone; and I went back and I checked my wife again; and now I was
—I don't know. I assume I was hoping I hadn't seen what I had
seen or I'd—or I was starting to think more like a doctor. So,
I went back and I checked for pulses. You know, carotid pulses
and stuff; and I—there was no pulse on my wife, and I was—
I felt I was getting sick to my stomach and I was short of breath,
and I was dizzy and my teeth were chattering 'cause I was cold.
And so I didn't know if I was going—I assume I was going into
shock because I was so cold. That's one of the symptoms of shock;
you start getting shaking chills. So, I got down on all fours;
and I was breathing for a while. Then I realized that I had
talked to the operator and nothing really had happened with her.
But in any case, when I went back to check my wife, I then went to
check the kids. And a couple time I had to—thinking that I was
going into shock and not being able to breathe. Now I—you know,
when I look back, of course, it's merely a symptom, that shortness
of breath. It isn't—you weren't really that bad, but that's
what happens when you get a pneumothorax. You—you think you
can't breathe. And I had to get down on my hands and knees and
breathe for a while, and then I went in and checked the kids and
checked their pulses and stuff. And—I don't know if it was the
first time I checked them or the second time I checked them, to
tell you the truth; but I had all—you know; blood on my hands
and I had little cuts in here and in here, and my head hurt. So,
when I reached up to feel my head, you know my hand was bloody.
And so I—I think it was the second circuit 'cause it—by that
time, I was—I was thinking better, I thought. And I went into
that—I went into the bathroom right there and looked in the
mirror and didn't—nothing looked wrong. I mean there wasn't
really even a cut or anything. So, I—then went out in the hall.

I couldn't breathe, so I was on my hands and knees in the hall, and I—and it kept hitting me that really nothing had been solved when I called the operator. And so I went in and—this was in the—you know, in the middle of the hallway there. And I went the other way. I went into the kitchen, picked up that phone and the operator was on the line. My other phone had never been hung up. And she was still on the line, and she said, "Is this Captain MacDonald?" I said, "Yes, it is." And she said, "Just a minute." And there was some dial tones and stuff and then the Sergeant came on. And he said, "Can I help you?" So, I told him that I needed a doctor and an ambulance and that some people had been stabbed, and that I thought I was going to die. And he said, "They'll be right there." So, I left the phone; and I remember going back to look again. And the next thing I knew, an M.P was giving me mouth-to-mouth respiration next to—next to my wife. Now, I remember I was—I don't know if it was the first or second trip into the bedroom to see my wife—but I saw that the back door was open; but that's immaterial, I guess. That's it.

On August 29, 1979, MacDonald was convicted of murder.

As of this writing, he has been denied parole several times. He remains in prison serving a life sentence.

Conclusion

The interviewer's task is not completed until the interview is properly documented. With most investigative interviews, the interviewer will need a recorded statement from the interviewee. Ideally, the interview itself and the interviewee's statement should be audiorecorded or videorecorded. In contrast, with helping interviews, documentation will consist of the interviewer's own notes about what transpired during the interview session.

In most investigative interviews, a statement written by the interviewee is required. This signed and dated statement is evidence that the interviewee voluntarily provided the information. The statement needs to be completed carefully in the interviewee's own words, but following procedures that protect the statement and its writer. The interviewer needs to be prepared for reluctance on the part of the interviewee to write a statement, and plan accordingly.

The recent addition of IDA provides another reason to have the interviewee's own words written down or recorded and transcribed. While its use is as an investigative tool only, it provides another technique to help interviewers get valid and complete information.

ALLEN'S WORLD

Dr. Lord and I agree a lot about note taking and probing. I have encountered some problems with my notes and tapes when confronted by prosecutors and other attorneys, so I have taken steps to make my interviews bulletproof.

Regardless of what you read next, students should realize that attempting to introduce notes, tapes, videos, affidavits, or other documents into evidence is a highly complex legal matter.

Sometimes, a dying declaration is admissible. Let us say a police officer gets to the scene of a shooting. There is a man lying in the street, bleeding profusely.

"Who shot you?" the policeman asks.

The man names his assailant and then he dies. The naming of the assailant, as much information as the officer can obtain, could be considered the *dying declaration.*

Can that identification be used in court? That is impossible to answer. Too many variables.

Let us say the police are conducting a field interview. They are talking to a man near the scene of the robbery. He has not been given his *Miranda* warning because he is not a suspect.

For some reason, not provoked by the police, the man just blurts out, "I robbed that store. I needed the money for my baby's doctor."

The blurting out of information is sometimes called an *excited utterance.*

Can that utterance by used in court? That is impossible to answer. Again, too many variables.

Students must be aware, regardless of how careful they are in taking notes, transcribing, taping, and so forth, that information may not always be allowed into evidence.

I write out all my notes, not the witness. I read them back and ask for corrections. I get the witness to sign and date each page of my notes, including any corrections.

If the interview is taped, I begin with the same information Dr. Lord suggests. I add the name of the attorney I am working for. I verify with the witness that I have identified myself as a private investigator. I tell the witness who I am working for and that my notes will be turned over to the attorney.

I ask the person to confirm that they are aware—if I am not recording it secretly—the interview is being taped, and I am doing so with his or her consent.

If there is someone else present during the interview, I list the names of those folks at the beginning of the tape. I ask each person to state his or her name for the record and who each is. That makes it easier, for instance, if the tape is being transcribed, for the person doing the transcription to know who all of the players are and to know their voices so the transcriber can properly attribute statements.

I have had tapes challenged in court. One time, during a proffer, the witness— believe it or not—denied talking to me. The first time that happened, the judge ruled the tape inadmissible. (A *proffer* is an offer by an attorney to a judge to determine if a document or testimony can be put into evidence. This always happens outside the hearing of the jury for the very reason that if the judge decides it cannot be entered into evidence, the jury is not supposed to hear the testimony or read the document.)

So, now the attorneys I work for often create an affidavit based on my interview notes or tapes. I swear the tape is authentic, it has not been doctored, it had not been turned off during the interview, and it is a complete and accurate recording of the interview. If possible, I take a copy of the affidavit to the interviewee and get him or her to sign it. I sign the affidavit before a notary. Since adopting that practice, none of my tapes have been ruled inadmissible.

I keep my probe notes separate from my interview notes. This avoids changes or corrections to the interview notes. I destroy my probe sheet.

Often, I turn off the tape or stop taking notes just to get the witness to relax. I chitchat, and it is amazing what they will say once they no longer feel threatened.

Every time I turn off the tape, the machine makes a little click. When I turn it back on, it makes a little click. So, anyone listening to the tape will know what happened when the tape was off; I say that I am turning it off so the interviewer can get a drink or use the rest room. When I turn the tape back on, I record the fact that no questions were asked and/or answered while the tape was off.

As soon as I leave, I write the chitchat notes and submit a report under separate cover. In that report, I also assess the witness's demeanor. Was something said in jest? Was the witness angry?

Remember notes, no matter how accurate, cannot reflect the tenor or tone or emotions. Was the witness crying, joking, or threatening?

For example, after the discussion of a case with the district attorney (DA), it seemed likely that an innocent man had been convicted and sent to prison for life. The DA said to me, "That's the kind of prosecutor I want. Anybody can convict a guilty person."

The DA was joking. We both knew it. We both laughed. The words themselves, however, could have created a controversy for the DA. (The DA agreed to give the convicted man a polygraph. If the man passed, the DA promised, he would ask the judge to set aside the original verdict, request a new trial, and dismiss the charges. The convicted man failed the polygraph. He is serving his life sentence.)

One interesting fact about notes reports, tapes, and statements is there is a tenet in law called "the work-product doctrine." Simply stated, it means that any work done by lawyers or the defense team is private and cannot be forced into disclosure.

PART

Special Areas

Interviewing Victims: Adults and Children

OBJECTIVES   Upon completion of this chapter, students should be able to

1. Incorporate effective interviewing techniques into the specific needs of victims
2. Understand and include the developmental issues of children in preparing and conducting interviews

Introduction

Victims of crime have been traumatized no matter the crime; however, victims of personal crimes such as assaults, armed robbery, and sexual assaults have additional issues that must be addressed.

Children who are victimized experience much of the same emotional responses as adults, but due to their chronological, emotional, and cognitive levels of maturity, they will perceive and cope with the trauma differently. Whether the child or adolescent is the victim of a property or personal crime, the interviewer must be sensitive to the individual's specific needs, what issues are to be addressed, and what motivates each individual. There must be a balance between the steps the interviewer takes to gather the necessary information for a complete, relevant, and valid interview, and the emotional needs that these young victims must have met after being traumatized.

Interviewing Adult Victims

Trauma is defined as an abrupt disruption in an individual's ordinary daily experience that causes loss of control over the body and may be perceived as objectification of the body (Holmberg, 2004). Victims often attempt to disassociate from the incident and whatever may have been done to them during the traumatic occurrence. Traumatized individuals become helpless because they now experience the world as unpredictable, threatening, and assaulting.

The traumatic event is foreign, an experience outside the range of the victims' ordinary lives, so they have little immediate coping resources. This lack of resources leads to distortion of the event or the banishment from consciousness. If the victim's mind cannot deal with the event, it will attempt to either repress it or forget about it. The perceived helplessness is often accompanied by pain and fear that springs from the sense of having little or no control over what happened to them. What will keep it from happening to them again? Often the police and others will hear the victims making responses such as," If only I had ____." The phrase may be completed with all sorts of protective comments, e.g. "locked the door," "realized that he was a bad person," or "called the police." Such phrases should not be taken as the perpetrator's lack of responsibility, but rather that the victims are attempting to reassure themselves that they will not be victims again. If victims can reassure themselves that they can prevent future crimes against themselves, they will feel safer, but these protective comments may make them sound culpable.

This stress-reaction process that the victims put themselves through demands that they receive continuous interaction and feedback between cognitive and emotional appraisal of the strain and its meaning for them. Interviewers must be receptive and provide space for victims to ventilate psychologically loaded issues in the interview. The interviewer needs to be trained to recognize when psychological help is needed by the victims.

These issues that the victims ventilate are not necessarily investigative facts; however, as the information is reviewed through talking, the facts might lead to additional critical information. Because the experience of these events causes an imbalance between the perceived demands of the events and the perceived resources at the victims' disposal, the victims feel overwhelmed. If allowed to ventilate, victims often see their situations through a less-anxious prism.

Chapter 4 emphasized rapport building, and its importance re-emerges with even greater vigor in this chapter. If developed correctly, rapport building, which includes finding common ground between the victim and the interviewer, showing empathy, and actively listening, makes it easier for victims to provide sensitive information because they believe the interviewers care about them.

As victims talk about the incident, they are ventilating their feelings. These feelings are particularly important during the first 48 hours after a crisis. Given the level of trauma and the victims' coping resources, the victims may develop acute stress disorder (ASD). Symptoms of ASD include the re-experience of the trauma, avoidance of reminders of the trauma, anxiety, hypervigilance, and impairment in social and occupational functioning (American Psychiatric Association, 1996). The victims may isolate themselves from their friends and family and any reminder of the event. It is critical that crime victims be interviewed within 2 days and that critical stress debriefing begins as soon as possible. If these symptoms continue more than a month, then the victims are considered to be suffering from posttraumatic stress disorder (Holmberg, 2004).

The victims quite likely will be suffering from ASD while being interviewed. Because victims often are traumatized by their experience, it is important for the interviewer to be sensitive to possible ASD symptoms, especially avoidance of reminders of the trauma and lack of concentration. A common symptom of sexual assault discovered by one of the authors of this book was future denial of the crime. Early in Vivian Lord's career as an investigator, she conducted a preliminary interview of a rape victim. When she returned the next day for follow-up, the victim denied anything had happened to her. It required extensive time and counseling before that particular victim allowed the incident to resurface into her conscious mind. After the experience of working with that victim, Vivian Lord always included a cautionary statement in her interviews with victims. She would tell the victims that their subconscious mind might attempt to suppress the trauma, leading them to refuse follow-up interviews. Lord was never again refused a follow-up interview.

It is unknown if the cautionary statement Dr. Lord used prevented the future refusals to continue interviewing and/or denials of the offense. Using such cautionary statements is supported with the literature on "preparing and predicting for" the victim. In helping the victim understand the feelings that he or she might feel in the future and discussing means to cope with those feelings allows him or her to recover as much control of his or her life as possible. After the interview with the victim, the interviewer will further aid in the victim's preparation and future prediction by outlining with the victim the steps that need to be taken. It should be kept simple:

> "Molly, I know you must be exhausted and just want to be able to shower and get some sleep. Once your friend brings you a change of clothes so we can collect what you are wearing for evidence, then we have collected all of the physical evidence from you. You will be able to leave with your friend, or we will be glad to take you home."
>
> "I want you to realize that it might take you a little time to get back to your normal routine. If possible, give yourself a little time. Will your friend be able to stay with you? When would be convenient for us to meet with you tomorrow? I don't want to burden you with any more questions or procedures tonight. Tomorrow will be soon enough to go over with you what needs to be done next to help with the investigation and help you recover."

Ulf Holmberg (2004) states that victims perceive attitudes of interviewers as characterized by either dominance or humanity. The dominance approach is distinguished by impatience, aggression, deprecation, condemnation, and brusqueness. The humanitarian approach is typified by cooperation, accommodation, empathy, positivity, and helpfulness. Holmberg concluded that most police interviews are marked by dominance. While dominance may (or may not) be an effective approach with suspects, it is not constructive with victims. The humanitarian interviewing style promotes crime victims' feelings of being respected and encourages them to provide all the information from painful events. If the interview is conducted using the humanitarian approach, it is more likely to aid the victim in overcoming the experience.

Even so, the interviewer should be familiar with community resources and be prepared to make referrals for professional help. ASD in victims often needs additional treatment, and the interviewer will want to be sure the treatment professionals work with the investigative team.

Familiarity with community resources should extend to previous introductions and meetings with victim advocates and resources. The author, Vivian Lord, served on the board of a rape crisis center early in her career. The advocates realized that she wanted to help victims as much as she wanted to get evidence that would lead to the arrest of assailants. They worked together to support the victims and to keep them strong for court when the perpetrators went on trial.

The interviewer has additional responsibilities to the victims besides just gathering information. Allowing victims to ventilate their initial reactions to the crimes helps the gathering of critical information about the incident. The interviewer also needs to be able to identify when victims are experiencing ASD and to refer them to treatment professionals. Together the professionals will keep the victims emotionally strong for the necessary criminal justice proceedings, but also will help them regain the emotionally stable lives they possessed before the crime. Chapter 11 expands the readers' knowledge on crisis and crisis intervention.

EXERCISE 7.1

Interviewing Victims

Activity: In role plays as victims and interviewers, students practice the use of skills in which they were trained from previous chapters.

Purpose: To integrate and apply basic interviewing skills to the special needs of victims.

Students should be divided into groups of three: speaker, listener, and observer. Because these interviews should be somewhat extensive, a class period should be dedicated to allow each person to be the interviewer at least once. The students as victims should be instructed to improvise, but provide information as they feel comfortable with the interviewer's approach. The following are some potential offenses that will not be too sensitive for students:

1. A victim reports theft from auto
2. A homeowner reports vandalism
3. A shop owner reports a break-in
4. An individual reports a simple assault in which he was slapped by an old girlfriend or boyfriend
5. A business owner reports a building break-in

After completing the interview, the interviewer reads his or her notes of the crime. The victim and observer complete the form below. Each should then discuss what they reported. The interviewer should be given an opportunity to discuss what he or she believes went well and what could have been done differently.

Criteria	Yes	No
1. Eye contact was maintained without gazing or staring.		
2. Body posture was appropriate (relaxed, slightly leaning forward).		
3. He or she made me feel comfortable and relaxed.		
4. By the use of probes, he or she made me really provide complete answers.		
5. He or she seemed to be genuinely interested in me.		
6. He or she delivered questions without hesitations.		
7. He or she often asked for clarification and often paraphrased.		
8. He or she accurately reflected my feelings.		
9. I felt that I could tell him or her just about anything.		
10. Rating of his or her reported accuracy of my report of the incident between 1 and 10 with 10 most accurate.		

Interviewing Children

Interviewing children follows the same principles as have been discussed in previous chapters. The child must feel accepted and not judged. In counseling situations, it is critical that self-determination and confidentiality apply.

Children usually do not initiate complaints or the need for assistance. They primarily will be victims or witnesses, who tell what has happened to them, to an adult, or a friend. Sometimes adults in caretaker roles notice a change in the child and persuade the child to tell them what happened.

The initial interview by this caretaker may have an impact on what the child tells the professional interviewer. If the caretaker becomes hysterical, the child may shut down and refuse to talk further about the victimization. Or, the caretaker may start asking leading questions, and the child feels compelled to say some actions occurred that did not. The interviewer needs to approach child victimization carefully and with an open mind.

The major difference between children and adults is the need to consider the children's developmental levels. Children communicate and perceive the world differently at different ages.

**Children's
Developmental
Levels**

Children across a wide range of ages will convey psychological distress through behavioral disturbances that may not be age appropriate. In fact, psychological distress often causes children to regress to an earlier emotional age that is demonstrated through a variety of behaviors. Bodily functions such as eating, sleeping, bowel–bladder control, speech, and motor functions may be disrupted such that the child may have problems sleeping and eating. They might wet their beds. Victimized children often suffer from memory loss and have problems learning in school. They may exhibit affective behavior such as fear, anxiety, depressive symptoms, hyperactivity, uncontrollable crying, and separation anxiety.

Erik Erikson (1968) describes stages that detail children's social and emotional development. Erikson argues that at each stage children are confronted by a psychosocial crisis that must be met before advancing to the next stage. If children experience trauma, they face an additional situational crisis that may cause regression to an earlier developmental level or delay in moving beyond the level they were in when the trauma occurred.

Erikson defines the following levels:

- Infancy (0 to 2): The child develops feelings of trust and security if nurtured and loved. A crisis during these early years may lead to insecurity and mistrust—especially if adults known by the child are the perpetrators or do not intervene to protect the child.
- Childhood (2 to 4): Children gain control and self-confidence that lead to autonomy, the desire to do things for themselves. When children experience trauma, their parents may become overly protective, delaying their children's' ability to achieve autonomy. If the trauma is caused by known adults, the children's self-confidence and desire for independence may be retarded.
- Play age (5 to 7): Learning initiative is the primary task for children as they begin school. They learn to imagine, to cooperate, and to lead, as well as to follow. Their skills broaden through active play and fantasy. Crisis and/or poor parenting during this stage can lead to fearfulness, overdependence on adults, and restriction in imagination and play skills.
- School age (8 to 12): The school-age stage continues until middle school, during which time children learn how to master more formal skills, relate to their peers, follow rules in teamwork settings, and master learning in areas of reading, arithmetic, and social sciences. Increase in self-discipline becomes a growing requirement. Children who have mastered earlier stages of trust, autonomy, and initiative can learn industry; however, crises in children's lives can influence their optimism and increase feelings of inferiority and defeat.
- Adolescence (13 to 18): Identity is the primary task as children complete separation from their parents. The question, Who am I? is answered as a reflection from their peers, parents, and authority figures. Children experiment with different roles and anticipate achievement. Later in adolescence, the child obtains clear sexual identity and develops a set of values and beliefs. Crises during adolescence can create confusion and paralysis as they seek their identity.

- Young adult (18 to 22): Intimacy may be repressed indefinitely if the individual is traumatized during this period. Intimacy is especially an issue if the trauma is sexually related.

When children are victimized, it may be evident to professional caretakers, such as teachers, that a child may not be functioning at the normal developmental level, or they may notice a child who has regressed during the academic year. It often is only when a teacher makes such observations and questions the child that the victimization comes to light.

Children's Cognitive Factors
Children as young as 2 and 3 years old can recall information, although it will be sparse (Hynan, 1999; London, 2001). As children get older, their answers are more complete. No matter the age, their memory is accurate as long as there have not been purposeful or inadvertent actions taken to distort it.

Numerous studies have been conducted to examine the accuracy of children's memory. Suggestibility, how vulnerable or resistant are children's memories (or the children's testimony), is a major factor. Language usage, ability to distinguish fantasy from reality, and tendency to lie are other factors that will be discussed.

To test suggestibility, Pezdek and Roe (1997) examined the ease of "planting" memory for an event that did not occur in contrast to the ease of changing a memory or erasing memory. Four-year old children and 10-year-old children were exposed to situations in which they were shown pictures of different objects by an adult. While shown the pictures, the children were either touched for 10 seconds on the hand or the shoulder, or not touched. In other words, it was an appropriate touch, not a push on a part of the body that the children would have been taught to interpret as inappropriate touching. About 15 minutes after they were shown all the pictures, the children were each given a memory test. Along with questions about the objects shown, the children were also asked the following:

> When I showed you the picture of the [object] on the screen and asked you if you could see it, did I touch you?
> When I showed you the picture of the [object] on the screen and asked you if you could see it, did I touch you on your hand?
> When I showed you the picture of the [object] on the screen and asked you if you could see it, did I touch you on your shoulder? (Pezdek & Roe, 1997, p. 100)

The majority of children, when asked about the touch location that was different than the actual location, agreed that the location touched was different than what actually occurred. In other words, if the child had been touched on the hand, the child agreed that he or she had been touched on the shoulder also. On the other hand, the majority of children who were asked if they had been touched, but had not been touched at all, responded negatively; they stated that they had not been touched. The children's memory could not be erased nor could a new memory be planted. So most of the children, no matter their age, did not agree that they had been touched when they had not been touched, but they did agree to a variation in what actually occurred. The researchers concluded that it was relatively easy to change a child's memory, but much more difficult to plant a memory or to erase a memory. The age of the child was not significant.

Researchers (Garven, Wood, Malpass, & Shaw, 1998) studied effects of social influence and reinforcement on children's memory. Their research was conducted after the McMartin Preschool case (*People v. Buckey*, 1990) in which 7 teachers were accused of sexually abusing hundreds of children during a 10-year period. Most of the charges were dropped after a 7-year investigation that resulted in one of the most expensive trials in California's history. Most of the investigation centered around police and social workers' interviews of the children and has been thoroughly critiqued. The interviewers used methods that have been labeled *social-incentive questioning*. These methods include suggestive questions, social influence, reinforcement, and removal from direct experience.

Suggestive questions, also called *leading questions*, are worded in such a way that the interviewee, even a child, realizes the answer the interviewer wants. The question always includes the answer the interviewer is expecting, for example, "The man had a mustache, didn't he?" The child realizes that the interviewer wants a positive response: yes, the man had a mustache.

Social-influence questions attempt to convince the interviewee that other interviewees gave a specific response: "When we interviewed Timmy and Mary, they said that Mr. Green would take the children into the woods and spank them. Is this what you saw also?"

Providing children with praise when they respond with one particular type of answer and scolding with another type of response is reinforcing the child, or even the adult, to answer based on the interviewer's reinforcement: "What a great response. You are so smart!"

Garven and colleagues' research (1998) begins with a young adult male wearing a big, brightly colored hat, introduced as Manny to preschool children. Manny told the children a story and then put a sticker with a character from the story on the back of each child's hand. Manny also gave each child a cupcake with a napkin that also had designs from the story. Manny then told them good-bye.

The children were divided into two groups: those who would be interviewed by different types of social-incentive questioning as used by interviewers in the McMartin Preschool case previously described; and a group of control subjects who received general questioning conditions.

Researchers interviewed each child separately a week after the storytelling session. After developing rapport, researchers asked all of the children eight misleading things that Manny did not do and four things that he did do. Children, no matter the age, were more likely to provide false allegations if social-incentive conditions were used rather than control conditions. These misleading allegations included bad behavior by Manny, one involved touching and a secret. The misspoken comments happened fairly soon into the interview, and the children became more acquiescent as the interview went on. So once the interviewer began to use one of the social-incentive conditions such as suggestive questioning, social influence, reinforcement, or removal from direct contact, the children became more agreeable to the misleading allegations as the interview continued.

Based on these social-incentive conditions, Garven and his colleagues (1998) developed an acronym SIRR: suggestive questions, social influence, reinforcement, and removal from direct experience. While suggestive, or leading, questions are the best known, the researchers concluded that social-influence questions have the most impact. Other people—telling children that some people had provided specific information—were the social influence, or peer pressure, that had the most impact. Use of positive and negative reinforcement also was found to be particularly effective. For example, positive reinforcement such as telling children to be smart or helpful and negative reinforcement such as rebuking children by stating that their answers are inadequate are powerful. Finally, removal from the direct experience, for example, the use of puppets to ask children questions as a form of make believe, sometimes led to false allegations.

Preschool-age children may not realize that interviews are strictly focused on fact. If play is included as part of rapport building and interviewing, there is an added risk of fantasy. For clarification, children should be asked if the information they are providing is based on their direct experience or if it is the result of what someone told them (Hynan, 1999).

Repressed Memory   Events, especially traumatic events, that have been lost to individuals' conscious memory for extended periods of time—years—continues to be controversial. Sexual-abuse victims who have repressed memories of the abuse and later talk about it in therapy are becoming more frequent. Given what is known about acute stress and posttraumatic stress disorders today, there is increasing evidence of stressful events becoming deeply entrenched in victims' memories (Hynan, 1999).

While memory studies continue to be controversial, there is general agreement that older children and adults cannot recall events that occurred before approximately 2 or 3 years of age (Hynan, 1999).

Language Development   Cognitively, another problem for children is their inability to understand the language of the interviewers. As part of this problem, the children do not state their lack of comprehension, but rather attempt a response they believe will please the adult interviewer, or they refuse to answer. Also, there is no point for the interviewer to ask, "Do you understand?" Often children are motivated to please and so they will say yes. It is better to ask the children to explain what they have told the interviewer.

London (2001) describes some specific confusing terms. Children may say no to a simple question, "Did he touch you?" even though they have said that "He put his fingers inside me." Young children find words related to touch confusing. Also, when asked if they remember an event, young children believe the event has to be forgotten first. They may say no because they always remembered the event.

Children do not understand emotional concepts until at least age 8, and even to age 14, children may have different conceptual understandings for emotions. So if asked, "How did that make you feel?" children may give an answer that sounds inappropriate for their experiences.

Once in the courtroom, numerous legal terms confuse children. This is understandable, since adults not familiar with the legal world are also easily confused when the questions include legal terms. A brief discussion at the beginning of interviews helps educate children about how to cope with questions. Children should be encouraged to ask for clarification of the meaning of any words they do not understand and should feel comfortable to state when they are confused by a question or do not know the answer. The interviewer should explain that he or she does not know the correct answer. The child is asked questions because the interviewer does not know the answers, and perhaps the child has helpful information.

Social Factors   Children are highly suggestible. They are motivated to please authority figures such as police officers and teachers. This desire to please can lead to productive responses but can also result in misleading statements. If the interviewer's tone of voice, facial expressions, and specific word usage conveys to the child critical connotations about another individual, it may encourage the child to make negative remarks.

Repeated questioning may lead children to agree with false allegations of abuse. While in some cases, repeated interviews help children add new details, there is a greater danger that children change their responses. The children believe that adults are continuing to question them because they are giving the wrong answers. When children are repeatedly interviewed by different professionals, the interviewers should explain to the children that they are not aware of the information that the children gave in previous interviews and would like to learn about it (London, 2001).

Use of anatomically detailed materials may lead to the child's expectation that the interviewer is looking for statements of a sexual nature and may influence the child's response. These materials should be introduced carefully. Specialized training is recommended for interviewers who will be using anatomically correct dolls.

Children at any age may lie. As with adults, children will lie to protect themselves or others and to avoid embarrassment. Interviewers should approach interviews with children in a positive fashion, communicating their desire to gain information without conveying any judgment about what children tell them. The more objective, but caring, demeanor the interviewers project, the more likely they can prevent fabrication by the young interviewees.

Building Rapport

Just as rapport building is important with adults, it is more likely to facilitate truthful and relatively complete information from children because they will feel comfortable. Interviewers should carefully consider the physical and verbal setting.

Physical Setting

The children's developmental level is a factor when considerations of interview locations are made. If more than one interview is going to be conducted with a child, the same setting should be used each time. To facilitate rapport building, the room should contain developmentally appropriate items such as electronic or board games for older children, and for younger children, items to facilitate symbolic play, e.g., doll houses and toys.

The child should be given time to become familiar with the interview environment and should be encouraged to ask questions. During the interview, the adult should sit on the floor or at a small table at the child's level when the child is school age or younger. Adolescents need a different type of room set up that is comfortable and informal, but more grown up.

Verbal Setting

Similar to adults, time should be spent connecting with the child to demonstrate personal interest. The interviewer should talk briefly about the child's friends, pets, or school.

The interviewer should identify himself or herself to establish legitimacy. This can be accomplished by the interviewer explaining that he or she often visits children to talk to them about things they like and do not like.

> "Hi, Joey. I'm Ms. Ellis. My job is to talk with children about things that hurt them or make them sad. Sometimes children I talk with are scared and don't know how to stop the hurting or the fear. I know not to push children, that when they are comfortable and feel they can trust me then they are more likely to let me know about the things that scare them."

During the early conversation, the interviewer can assess the child's level of sophistication and ability to understand concepts. With young children, the interviewer must assess the child's ability to distinguish between reality and fantasy.

Conducting the Interview

The play that began during rapport building is effective for the interview. Fantasy play using puppets, dolls, drawings, or a play telephone is effective for preschool. Elementary children are more interested in art supplies and action toys, while preadolescents and adolescents are more comfortable with direct interviewing.

Depending on the age of the children, the interviewer might begin by observing children playing with toys and games available to them. With proper training, the interviewer might be more directive by placing specifically chosen toys out for children. As the children play, the interviewer asks them about their play. To repeat an earlier caution, the interviewer needs to be careful when attempting to translate what children do in play to what has actually happened to them. Using play or drawings should be primarily used to clarify information the child has already disclosed.

Use of puppets or dolls is encouraged with young children; however, they must be used with caution. Rather than answering questions directly related to the incident, the children answer questions that are not directly experience related. If the interviewer asks the child to play with a doll house, and the child gives each doll a name of a family member, the dolls are still dolls. The interviewer must be careful in the interpretation of the doll daddy touching the doll baby as equivalent to the child's father touching the child inappropriately.

Researchers at the National Institute of Child Health and Human Development (NICHD) developed a protocol for extensively trained police detectives to use when interviewing suspected victims of abuse between the years of 1997 and 2000 (Pipe, Orbach, Lamb, Abbott, & Stewart, 2006). Pipe and colleagues evaluated the effectiveness of the NICHD protocol. The researchers found that charges against suspects were more likely to be filed for cases involving the use of the NICHD protocol than cases interviewed by the same detectives before receiving training in the protocol. More of the cases using NICHD protocol also led to guilty pleas and convictions of the perpetrator. The researchers concluded that improving the quality of children's interviews greatly increased the probability of cases of alleged sexual abuse being prosecuted (Pipe et al., 2006).

The NICHD protocol adapts forensic interviews to age-appropriate information requests and includes techniques that are effective in eliciting free-recall information from even young children. In the introductory phase, the interviewer clarifies the child's task. The child is to describe events in detail and to tell the truth. The child should admit lack of knowledge if he or she does not know, indicate when the question is not understood, and correct the interviewer when necessary. During the rapport stage, the goal is to create a relaxed, supportive environment and to connect with the child. Nonsuggestive prompts are used to target the event that is the focus of the interview. The free-recall phase begins with an open invitation: "Tell me everything that happened from the beginning to the end as best as you can remember." The open phase is followed with open-ended prompts such as "Then what happened?" Later prompts are still open: "Earlier you mentioned (person–object–action). Tell me everything about that." Only after open-ended questioning is exhausted does the interviewer proceed to directive questions such as "Where were you when that happened?" If crucial details are still missing at the end of the interview, the interviewer may ask limited force-choice questions such as those that can be answered with yes or no (Pipe et al., 2006, pp. 23–24).

Formulating Questions

Question development for children is difficult. If the interviewer attempts to ask open-ended questions, the child is likely to provide little response, omitting a great deal of relevant information. The children are more likely to answer open-ended questions about pleasant, noncontroversial areas of life first. If specific, close-ended questions are asked, the danger of asking leading questions increases, raising the probability that the children may include incorrect information.

The interviewer should begin with open-ended questions when possible, but probe more directly than with adult interviews. Try to let the child tell what happened before asking specific questions. Contextual statements are an important tool to explain why specific questions need to be asked: "Mary, I am sorry if these questions make you uncomfortable. Please take your time and answer when you are able. I have to know these details in order to find additional information to support what you are telling me. It is called corroboration, and we have to do it with every case."

The younger the child, the more difficult it is to sustain and articulate open-ended questions. Limited vocabulary may result in sparse responses. The dilemma is how to ask questions so the interviewer receives complete responses, but does not ask leading questions. The number of questions should be limited and carefully worded so the interviewer can be sure the child understands and has the opportunity to elaborate and offer other information spontaneously. As noted earlier, it is important to develop a climate in which the child will feel comfortable stating that he or she does not understand if the question is confusing.

Maximizing Facilitators— Minimizing Inhibitors

To minimize ego threat, begin with neutral questions. The interviewer should ask for reactions from the child about talking to the interviewer and about the interview taking place at the agency or hospital. Because, for children, talking to adults can add to their feeling of being punished, the interviewer should anticipate the child's concerns, including his or her fear of being punished for talking to the interviewer.

To maximize memory recognition, emphasize the importance of getting information from them: "Somebody who is concerned about you called me today to say you have a problem at home and need some help. What can you tell me so I can help you?"

Cognitive interviewing, discussed in Chapter 3, has been modified for child witnesses and, as with adults, has been found to elicit more accurate information than the standard police interview (Kebbell & Wagstaff, 1999; London, 2001). Having the child psychologically reconstruct the situation at the time of the event and report even what they may think is unimportant will add to the initial report by the child. The child should be encouraged to include information about all the parties involved, conversations among the parties, and the emotions portrayed by the parties.

To minimize forgetting, inquire about the time of day the events occurred. The time can be framed around constants in the child's life, e.g., before a specific TV program or holiday or bedtime. Until adolescence, children have difficulties with measurement of time. Requesting the time of day that an event occurred, especially if it was reoccurring, can usually only be pinned down contextually. In other words, the traumatic event needs to be linked to the normal events in the child's life. For example, the event occurred when the child returned home after school or during bath time.

The interviewer should relay a great deal of empathy. The child may say that she or he has forgotten and does not want to talk about it: "It must be really hard to talk about . . ." The interviewer must be especially empathic to children by being sensitive to their discomfort: "I know these questions bother you, but you are doing a really good job." "Would you like to write it down instead?" "Take your time. I know it's hard to talk about it."

The interviewer also should be particularly sensitive to anything that could be perceived as a bribe or enticement, such as giving the child ice cream or another type of treat. Such enticements are regularly used by abusers to convince the children of their good intentions. It is best to restrain from providing treats.

Let the children use their own language. It will add to their comfort level and certainly appear genuine: "What happened after Papa Jones touched your wee-wee?"

EXERCISE 7.2

Using Children's Language in Sensitive Interviews

Activity: Brainstorming common words used for body parts.

Purpose: Because child physical and sexual abuse cases are common for child victims, interviewers need to be familiar with and comfortable using common children's words for different parts of the body.

The instructor divides the class into groups of six to eight students. The groups are instructed to list all the slang and proper words they know for female and male genitalia. Limit the time (10–15 minutes). The instructor calls time and then tells each group to count the words in each list. The group with the most words in each list is then to call out the words. Other groups will then add any additional words they might have listed.

It is important for the instructor to thoroughly process this exercise. Ask the students how they felt saying these words and then publically calling them out. They will probably mention embarrassment and discomfort. Ask them how they think children will feel talking to professionals who are strangers about these matters. Discuss the importance of empathy, expressing an understanding to children of their discomfort.

Because the accuracy of children's interviews will be questioned, it is imperative that the interviewer look for evidence from other sources that may collaborate or contradict information given in the interview (see Allen's World at the end of this chapter for examples). Psychological evaluation of the children is helpful in many of these cases.

Terminating the Interview

The interviewer should provide advance warning for the ending the interview: "You and I are going to talk until 9:00," or "I really appreciate how hard you have worked to try and remember all of the facts. I only have two more questions."

The child should be given the opportunity to ask any final questions. "What questions do you have? I will try hard to answer your questions about the interview or what happens next."

It is important also to prepare and predict; let them know what will happen next: "Next week when we meet, you can play with the same toys," or "I will come to your house tomorrow with Ms. Brown. I think you will like talking to her. Your mother has met her."

It is especially important for the interviewer to use some sort of ritual to end the session if there will be multiple interviews over several weeks. These rituals can be as simple as picking up toys, putting away a game, or putting pillows back on the couch.

Conclusion

All victims have special needs. Interviewers should be sensitive to possible stress reactions with which the victims are coping. Rapport building and empathic responses are critical for accurate information and the ability for victims to emotionally heal from the trauma.

Interviewing children requires special techniques, and interviewers need to be knowledgeable of children's developmental stages. Their cognitive, language, and social development is particularly relevant, and age must be taken into consideration. Knowledge about factors that influence the children's statements will improve the accuracy of the interviews, leading to young victims benefiting from help provided by the criminal justice system.

EXERCISE 7.3

Interviewing Children

Activity: Practicing interviewing children.

Purpose: Help students comprehend some of the difficulties in interviewing children.

Ask students if they have younger siblings, cousins, or nephews and nieces. There may be other opportunities for children interviews. For instance, if the academic institution has a child development center, it might be possible to get permission to interview children.

Have the students work in pairs so one student interviews the child, and the other observes. The child should only be interviewed once. The student should select a harmless topic, for example, what did the child do for his or her last birthday, vacation, and so on.

The student interviewing should prepare and ask open-ended questions much as the questions for his or her professional interview. The observer should take notes on some of the following points:

1. How well was the child able to respond? Make note of the age of the child.
2. Did the child understand the questions?
3. Did the child ask for any of the questions to be explained or ever answer "I don't know?"
4. How well did the interviewer keep to open-ended questions? Were any of the questions leading?
5. When the student probed, did the questions remain unbiased? How directed did the probes become?
6. If the child was not related or did not know the interviewer, how did the interviewer gain rapport?
7. Did the interviewer respond empathically?

ALLEN'S WORLD

As a private investigator working for criminal defense lawyers, I rarely interview crime victims or children. That is the job of law enforcement. My job is to interview the defendant's charged with the crime, to read the police reports and interview notes, and to look for ways to impeach the victims. My job working for criminal defense attorneys is to find collaborative or conflicting evidence. That is why, as Dr. Lord mentioned, it is important for the police detective to find evidence that supports or conflicts with the victim's statements.

Because North Carolina has a reciprocal-discovery rule, the attorneys I work for get the statements from the victims, witnesses, and crime lab reports about forensics.

Then I go to work. Following are some examples of what I do.

Mike was charged with first-degree murder. Two witnesses in the rental house where the murder was planned told the police they heard Mike planning the murder. They told police they were on the couch in the living room. Mike was in the kitchen with four other men. No doubt about it, the witnesses told police, they could hear Mike talking in the kitchen, and Mike was the ringleader. Mike was planning to kill somebody that night.

I talked to the rental agent, told him what I needed, and he agreed to meet me at the house and let me in. I took two people with me and a tape recorder. The house was long and rectangular. The living room was at one end, and there was a hallway maybe 60 feet long with bedrooms on each side. The kitchen was off to the side at the end of the long hallway.

I asked the two people I brought with me and the rental agent to stand in the kitchen and carry on a normal conversation. I made the assumption that *if* there were people in the kitchen planning a murder, they would not be talking loudly—hush hush, as in whispering.

I was standing where the couch was the night of the murder. I could not hear any conversation. I recorded the sounds of silence for my attorney. I then asked the three people in the kitchen to talk louder. More silence, which I recorded. I asked them to talk really loud. Again, more silence that I recorded.

The witnesses had to be lying. Along with some of my other findings, the attorney gave my tape to the assistant district attorney handling the case.

What was the outcome? The case was dismissed.

I said I rarely interview the victims, but in one instance my client was charged with armed robbery and assault. We had the victim's statement, and the woman was adamant in her identification of my client as the robber.

My attorney asked me to interview the victim. Much to my surprise, she agreed. Using the statement she gave to the police as the basis for my questions, I asked her to repeat what had happened. I wanted to match her memory to the statement and see if there were discrepancies.

In the police report, she said she had a $20 bill, and she thought $11 in singles. She repeated that exact amount in my interview. She was certain of the amount—positive. Well, the man who was arrested had a $10 bill in his pocket, and $11 in singles. Not a huge mistake, but a mistake nonetheless.

Curiously, there was no physical description of the assailant in her statement to the police. I asked her to describe the man: black, about 200 pounds, maybe 6 feet tall. Our client is black, about 280 pounds, maybe 6-foot-5 or 6-foot-6. Again, another mistake that might lead to her impeachment as a witness.

The victim told me the brand name of the shoes the robber was wearing. She was certain because she had a pair by the same manufacturer. Well, we checked the inventory report of what the man was wearing when he was booked and made to change into the mandatory orange jumpsuit handed out at most jails. His shoes were a different name brand.

As of this writing, this case has not gone to trial.

As for interviewing children, I cannot remember anytime in my 20 years as an investigator when I have interviewed a child—teens, yes; young adults, yes; but never a child.

When interviewing teens, I always talk to the parents first. Is it okay with them? If they hesitate, I invite them to sit in under one specific rule: they cannot prompt the teen or provide answers. Just silent sentinels to ensure their youngster is not badgered.

I always tape interviews with teens or young adults—safety for them and more safety for me in case they make allegations.

I do reinforce Dr. Lord's statements about interviewing children. They are so susceptible to suggestions; these types of interviews are delicate. For example, I rarely ask a child a leading question because the child will incorporate the information from the question into his or her memory, and from then on, the recollection is faulty.

However, as I do with adults, when children are involved, I get their statements from the police or social worker, and I can look for ways for my attorney to challenge their testimony. For example, my client, Chris, was charged with molesting a minor. They found each other on the Internet. Chris arranged to meet the young man in the parking lot of a county park. The minor told police that once he got into Chris's car, Chris fondled him. The young man said he went home, went to sleep, and told his parents the next day. That, to me, seemed suspect.

The young man said he knew the exact time he left his house for the rendezvous because he had been watching television. This supports Dr. Lord's guideline of getting children to remember an incident in the context of an event. Fortunately, Chris spent the afternoon at a friend's house. Chris left the friend's for the liaison with the young man and had to return to the friend's house because they had dinner plans for a fixed time. The friend confirmed the time Chris left and the time Chris returned. I had a timeline that could be checked.

I met Chris at his friend's house. We drove to the park. I told Chris to speed, run any light that he could, and take the most direct route. I filmed the outgoing trip and timed it. We turned around immediately and drove back to the friend's house with the same guidelines: speed, run the lights, take the most direct route. I filmed the return trip and timed it.

The victim's timetable was wrong. Not by a few seconds or minutes, but by more than 20 to 30 minutes. This meeting could not have taken place when the victim said it did.

Dr. Lord is dead on. When a child is a victim of a crime, his or her behavior changes. If not, the interviewer should be suspect of the information given by the child. I went to the alleged victim's school. He attended all his classes the day after the incident. He did not ask to see a counselor. He took a test and did well. His teachers did not notice any change in behavior. The youngster was on the junior varsity football team. His coach said he came to practice, no change in behavior.

All of this information was turned over to the prosecutor. The felony charges were dismissed. My client pled guilty to one count of a misdemeanor—no fine, no jail time, and no community service.

Something happened in that parking lot, but only Chris and the young man know for sure what took place.

(Just as an aside, it is curious to me how my clients react to the investigations I do, and how it affects their case. In this case I was thinking, "The attorney and I saved Chris from spending years in prison. Chris would be appreciative. His attorney and I did a great job for Chris." I was so wrong. Chris was angry that he was not exonerated. Chris was furious that his name would now be posted in the courthouse as a convicted sex offender. And that if he moved to another jurisdiction, he would have to report to the local officials to let them know of his presence.)

(Chris owed his attorney a lot of money. It took years of pressuring Chris before he finally paid his invoice. Go figure.)

Arnold (not his real name) was originally charged with several felony counts of molesting his two granddaughters. We had their statements given to us by the police and the psychologists who interviewed them. Part of the extended family included two other young girls; let us call them Nan and Jan. I interviewed their mother after the initial charges were filed against Arnold. Both Nan and Jan told their mother, which she repeated to me, that Arnold had never molested them.

Months later, the police charged Arnold also with molesting Nan and Jan. This was after the second set of girls talked with the first set of girls. We had statements from Nan and Jan. Much of what they said happened was identical to what the first set of girls reported. It was impossible to tell what, if anything, actually happened to Nan and Jan. In fact, the younger girl, Jan, never gave a statement. She told her older sister, Nan, what Arnold did, and it was the older sister who spoke on behalf of her younger sister.

All four girls were interviewed by counselors and the police. None of their statements matched. In the statements, the girls said at times Arnold molested them in a workshop behind his house, where he took them to make bird houses. Arnold was a shade tree auto mechanic. There was not a single piece of woodworking equipment in the workshop—no wood chips on the floor and no pieces of wood to make bird houses.

The girls described a chair Arnold made them sit in. There was, in fact, a chair similar to the one described by the girls in the workshop. However, it was covered with dust and piled high with stuff. It looked like it had not been used in centuries.

The girls told the police that other times Arnold molested them in a bedroom off the living room. I went to the house and mapped and measured the entire floor plan. The bedroom in question was less than 10 feet from the living room and not an ideal place to molest children when other adults were nearby.

The girls said Arnold gave them candy from a jar in the living room. I could not find that jar. The girls told police they used to work with Arnold in his office. Arnold never had an office. They told police Arnold kept food for them in a refrigerator in one of the bedrooms. The only refrigerator in the house was in the kitchen. Arnold's wife confirmed that none of the items described by the children ever existed.

The attorney I worked for, a former prosecutor, knew that despite the evidence suggesting the children were embellishing, cross-examining young girls is precarious. Any sign of badgering just elicits more sympathy from the jury.

My findings were given to the assistant district attorney prosecuting the case. The prosecutor offered to drop most of the felony charges if Arnold would plead to some minor offenses and serve no more than 24 months in jail.

Arnold reluctantly took the offering, insisting to this day he was pleading to crimes he did not commit. Arnold spent his two-year sentence in a minimum-custody facility near the Blue Ridge Mountains. He spent his time reading the Bible and walking the perimeter fence for hours. He is now back home with his wife, most likely sitting on a swing in his backyard and fixing cars for friends.

While my work often helps guilty people avoid jail time, Arnold's is the first case I have worked in which I believe an innocent man went to prison. My interviewing skills could not prevent Arnold's incarceration.

8 CHAPTER

Cultural Differences

OBJECTIVES   Upon completion of this chapter, the student should be able to

1. Understand general differences among cultures
2. Develop empathy toward individuals from different cultures
3. Be aware of nonverbal as well as verbal differences
4. Appreciate the importance of pursuing additional knowledge on specific cultures and of learning another language

Introduction

Interviews in the criminal justice system are increasingly intercultural. The criminal justice system is attracting more diverse professionals, and its clientele are becoming more international. It is important to be continually reminded that the goal of the interviewer is to obtain a complete, relevant, and valid interview. Accept the fact that the potential barriers of culture and language differences add vast challenges to meeting that goal.

Intercultural communication is the transmission of a message from a member of one culture to be understood by a person from another culture in hopes of creating a shared meaning (Lustig & Koester, 1999; Samovar & Porter, 1991). The message is grounded in the sender's culture.

What is a culture? Culture is a system of meaning, often translated through a collection of symbols for a group of people. The oral and written language of a group of people is a major part of the collection of symbols. As discussed in Chapter 1, words are symbols that represent concrete objects and abstract thoughts of people. Knowledge is formed and shaped by the way each culture of people processes information and creates its reality (Gudykunst, 2003). Words as concrete as *tree* and as abstract as *justice* are interpreted slightly different based on the experiences of different cultures.

Cultures are made up of the lore, myths, traditions, and history of a group of people. Individuals generate messages while adhering to a variety of unconscious constraints that they learn throughout their lives from their culture. Children are taught by people who are responsible for socializing them. The children are taught explanations about the natural and human events around them. These explanations include beliefs, values, and norms, which affect behavior and development of guidelines and constraints about what things mean, what is important, and what is permissible (Lustig & Koester, 1999). These constraints influence the manner in which messages are constructed and the individuals' conversational styles (Gudykunst, 2003). Linguists have developed different means to categorize these cultural differences.

Intercultural communication is particularly demanding because the culturally different individuals who are communicating have little common information in general.

The first important issue to a successful interview with a person from a different culture is finding a common language that both parties can use to work effectively. One of the two parties probably must use a second language, creating cognitive strain for both parties. Because the first-language speaker is likely to dominate the interactions, he or she must work to create an environment in which it is possible to check for understanding. The second-language user is contending with multiple demands of interpreting and speaking in a language that is less understandable to him or her. Both parties must devote substantial amount of attention to the communications process to achieve an effective transfer of understanding (Pekerti & Thomas, 2003).

The criminal justice interviewer is expected to communicate with people from a variety of cultures and languages, often using translators. The more the interviewer can understand the differences among cultures, the more likely he or she can truly develop rapport and empathy with interviewees from other cultures.

This chapter provides an overview of the important concepts, differences among the major cultures, and techniques to be used for intercultural interviewing; however, the only genuine means of developing a deep understanding of any culture is through visiting other countries, learning another country's language, and attempting first hand to assimilate into a different culture.

For some cultural differences, there is no need to travel too far. Within the United States are strong regional cultures. For example, citizens who have lived in the southern states for three or more generations consider themselves southern rather than their original Irish or German heritage. Also, it is important to consider cultural implications of sexual orientation. Cultural aspects of the homosexual lifestyle must be considered with an open mind and sensitivity.

Dimension of Cultural Variability

Although each culture has unique components, there are commonalities that help categorize cultures. Cultures are classified in categories of individualism or collectivism, high and low context, horizontal or vertical, masculine or feminine, and/or monochromic or polychromic (Gudykunst, 2003; Lustig & Koester, 1999; Pekerti & Thomas, 2003; Samovar & Porter, 1991).

Individualism Versus Collectivism

Members of individualistic cultures are described as independent, competitive, time conscious, results oriented, impatient, and assertive. The behavior of people from individualistic cultures is aimed at individual goals with individual opinions, and they strive to be self-reliant in solving problems

In contrast, members of collectivistic cultures are team oriented, closely affiliated with their communities, and respectful of other group members. They are considered good listeners with indirect communication styles and an understanding of the importance of emotions (Shearer, 2005). In collectivistic cultures, people are dependent on others in their communities, and are concerned with emphasizing, strengthening, and smoothing interpersonal relationships. Their communal feelings toward each other inhibit any need to compete. For example, life on a kibbutz in Israel is dedicated to the proposition of one for all. The goals of all the people on these collective farms are all that matter to their members.

One way to understand collectivistic behavior is to watch a football game. While individual team members' successes help the team win, the goal is for the team to win. Remember, there is no *I* in team. Bowling on the Professional Bowling Association tour, on the other hand, is an example of individualistic behavior. Each member competes alone and against every other bowler. Whoever knocks down the most pins wins. Every bowler must adapt to lane conditions, such as too much oil, and spar with the psychological ploys of their competitors.

Individualistic people view themselves separate from nature and instead believe they must master the environment to be successful. They move quickly, always trying to keep up with the "moving river of time." People from collectivistic cultures perceive themselves as part of nature, adapting to the environment. Time is a still pool, moving slowly, if at all.

Individualistic cultures desire clarity in communication; direct requests are the most effective strategies to accomplish goals. Confrontation is acceptable and even encouraged under certain circumstances. Collectivistic cultures prefer to avoid confrontation and work through intermediaries when necessary. Collectivistic cultures are concerned with feelings and not imposing their will on others. "Saving face" is a critical component of their communication. (Gudykunst, 2003).

Wei Young sits quietly on the edge of the couch with his hands in his lap over a folder. Although he is sitting straight, his eyes are focused on the floor in front of his feet. With him is another man, who sits alertly, scanning the room. From time to time, the other man says a few Vietnamese words to Wei Young, who quietly responds.

The interviewer approaches the two men and asks, "Wei Young?" Wei Young looks up quickly and then back at the floor. The other man quickly rises and approaches the interviewer, "My name is Tian Lu. Wei Young has asked me to come with him and explain his circumstances."

The interviewer looks at Wei Young, then at Tian Lu, back at Wei Young, and then says, "I will need for Wei Young to sign some forms authorizing you to talk about this circumstance. Perhaps Wei Young will answer a few questions for me after you talk?"

Tian Lu spoke quietly to Wei Young, who glanced shyly at the interviewer and then responded to Tian Lu. Tian Lu said, "I will explain the forms to Wei Young, and then he will agree to sign the forms. His English is not very good, but he will try to answer questions. I will probably need to translate your questions and his answers."

High Context and Low Context	Pekerti and Thomas (2003) compared intercultural communications between cultures they classify as high context and low context. The Asian culture used in their study was high context. The Pakeha culture used in their study was low context.

In high-context cultures, a great deal of the message is implicit; words convey only a small part of the message with gaps filled based on knowledge about the speaker, setting, or other contextual cues. High-context communicators also are usually part of collectivistic cultures that value harmony and focus on interdependence among people. The social esteem of others (face) is the main emphasis. The interviewer must realize in these situations that the parties accommodate and change their opinions rather than risk disharmony. Inconsistencies must be probed.

The other communicators were Pakeha, a culture of low-context communicators. Their messages are mainly conveyed by spoken words. Low-context communicators are usually part of individualistic cultures. The Pakeha exhibit communication behavior that is directed toward task accomplishment. They initiate action; express their opinions confidently; exhibit dominant, aggressive, and regulating behaviors; and produce logical arguments to achieve their tasks.

Pekerti and Thomas concluded that interacting with a member of a different culture enhanced individuals' dominant communication styles. Rather than adjusting their behavior in the direction of their counterpart as recommended, each communicator exhibited higher intensity of their dominant behavior. The Pakeha communicators became even more aggressive and dominant toward the Asians. The Asians quickly reflected the Pakeha's opinions rather than their own.

Without explicit situation cues to facilitate the conversation, the communicators begin to rely on their culturally based norms for behavior. In the presence of a culturally different individual, each may feel responsible for representing his or her own culture and displaying culture-specific behavior. Depending on indirect communication and nonverbal cues, the Asian communicators carefully observed the Pakeha's cues and moved away from anything that might appear to be confrontational. The Pakeha provided logical arguments for their rationale of the methods to accomplish the tasks and assumed the Asian communicators agreed.

In situations in which interviewers must talk to high-context communicators, they must keep in mind that high-context communicators are likely to appear as if they are agreeing so that harmony can be maintained. In settings such as probation or treatment-intervention programs, specific conditions, such as getting a job or finishing school, are likely to be established. The individual may not have the necessary means such as transportation or child care to achieve the conditions, but would be uncomfortable describing barriers to the objectives. They will just agree with the person in authority, all the while knowing they cannot meet the conditions.

EXERCISE 8.1

Working as a Team

Activity: Drawing houses as a dyad team.

Purpose: To provide students an opportunity to understand how to communicate and achieve a task without words.

Students are paired up, encouraging dissimilar teams, e.g., different race, ethnicity, gender, or ages. Each team is to have one piece of paper and one writing instrument such as a pen or pencil. They are given instructions to each grasp the pen/pencil with one person's fingers higher on it. They can be told they will be asked to draw two pictures, so they will switch places on the second picture. They are not to talk to each other at all.

The first picture is to be a house. Remind them not to talk. When everybody has completed the first house, tell them to switch places on the pen/pencil. The second picture is to be a Mongolian House. Remind them again not to talk.

To process this exercise, the instructor will ask the students how they made the decision of who was to be on the bottom or top of the writing instrument. How did they make decisions on what to draw for the first house? How did they make decisions about the second part of the exercise, in which more than likely nobody had ever seen a Mongolian House?

How do they think this applies to interviewing individuals who do not speak their language very well and are from a different country/culture?

Horizontal Versus Vertical Cultures

Ranking among people is emphasized more in some cultures than others. Individuals are not expected to elevate themselves around members of their own or equal groups in horizontal cultures. Equality is highly valued. In horizontal individualistic cultures like Sweden, individuals are expected to act as individuals, but also are not to stand out from others. Each person is expected to accomplish his or her own goals with everybody's successes considered equal.

In vertical cultures, individuals tend to see themselves as different from others, and equality is not valued highly. In a vertical-collectivistic culture, such as Korea, individuals conform to their in-groups but also are expected to stand out from their in-groups. For example, all students are to work hard at similar academic regimes; however, it is acceptable and expected for students to excel at different rates and to receive recognition for their higher achievements.

Higher achievement propels variation in ranking. Professionals such as teachers are highly regarded and addressed by their profession.

Perception of criminal justice professionals varies widely in respect and rank. For some cultures, the level of corruption in the criminal justice system may require communication of rank based on fear rather than respect. It is important to understand that even though the individual is living in the United States, his or her perception of the criminal justice system may be negative and based on experience of corruption and fear. The interviewer must find the means to build rapport and break down the interviewee's negative experiences.

"Lui Gail, my name is Sergeant Steve Spalding. I am in charge of the robbery and extortion unit for the Maple View Police Department. I would like you to know a little about me and my experience with these types of crimes. I would like for you to know also how we investigate these crimes and protect the victims."

"I graduated from State University fifteen years ago and began working for the Maple View Police Department soon after. After serving as a patrol officer for five years, I received training in basic investigations, interviewing, and diversity training with a concentration in the Asian cultures. Recently, I attended the FBI training program on extortion. I was assigned to the larceny investigation unit ten years ago and have been specializing in robberies and extortions for the past five years."

"I understand how much courage it takes to come forward with information about offenders who are preying on their own people. These offenders are threatening innocent hard-working folks who are trying to make a better life for their children. We have to figure out a way to work with you so that you can feel safe with providing us information. I understand that I can't just go up to these guys and arrest them for threatening you. They can bond out or call on others to retaliate against you or your family."

"But as you know these offenders are preying on lots of people. What I have become successful in doing over the past five years is keeping information confidential until I am able to compile sufficient information from a number of sources. In that way the offenders are unable to identify the sources. After talking to you, I need for you to identify just one other family. We take it slow. After a few weeks, I will approach them with a few careful questions. . . . "

In individualistic vertical cultures such as the United States, individuals are expected to act independently and stand out from others. People in the United States often do not place a value on equality, but instead they place a higher value on freedom. Students in the United States are expected to compete in a variety of areas, with creativity highly recognized.

Masculinity/ Femininity

Another means to categorize cultures is recognizing the difference in how gender roles are distributed within that culture. In masculine cultures, social gender roles are clearly distinct, while in feminine cultures, social gender roles overlap. In masculine cultures, members value performance, ambition, power, and assertiveness. Individuals organize information about other people based on their sex. Masculinity is predominant in Arabian countries, Austria, Germany, Italy, Japan, New Zealand, Switzerland, and most South American and Latin American cultures.

In feminine cultures, men and women engage in the valued behaviors of performance, ambition, power, assertiveness, tenderness, and caretaking relationships. Androgyny predominates in feminine cultures. Cultures that tend to be mainly feminine include Chile, Costa Rica, Denmark, eastern Africa, Finland, Netherlands, Portugal, and Sweden. The United States is slightly closer to feminine than masculine according to Gudykunst (2003); however, while the glass ceiling is changing in the United States, it remains in place in many aspects. How many presidents, CEOs, and members of the Supreme Court are or have been female? Is there equal pay between men and women doing the exact same job?

Interviewers always must be knowledgeable of the roles different members of families occupy when attempting to interview a family member alone or even with other family members around. In masculine cultures, requesting to interview a female member of the household by herself may be met with resistance, especially if the interviewer is male. The male adult in the household not only may want to remain during the interview, but also may want to respond for the female member. The interviewer must prepare before the interview on how the female interviewee can be approached so she responds to questions without her partner's influence.

Masculine cultures also may place some barriers for female interviewers. The criminal justice agency will have to make decisions on dealing with clients' refusals to speak to female interviewers. This inhibition can be an issue particularly in Middle Eastern cultures. Although individuals in other cultures are living in the United States, some cultural barriers, especially if they are religion-based are difficult to scale.

Perception of Time

Time is an important component of culture and communications. There are differences in how time is encoded, and different orientations toward temporally related situations. Cultures' organization of time is classified as either *monochromic* or *polychromic*. Monochromic cultures perceive time as stationary, so tasks should be done one at a time—in other words, no multitasking. Monochromic cultures emphasize schedules and promptness; time must not be wasted. Time is treated in most of these cultures' language as a noun. Time becomes an object that has substance and is countable (Gudykunst, 2003).

Polychromic people deal with time in less tangible terms. They have words for temporal cycles, but not the abstract notion of time. Time is a reoccurring event so people involved with an activity are important, not a schedule. Arabs, Greeks, American Indians, Mexicans, and some Africans are polychromic (Samovar & Porter, 1991).

Emphasis in the past, present, or future appears to follow different orientations than monochromic or polychromic. The past is emphasized in English, Chinese, American Indian languages, and Greek. On the other hand, Filipinos and Latin Americans live in the present with emphasis on living for the moment. Islamic cultures also live in the present: "the future belongs with Allah." The majority of U.S. citizens and those from countries whose cultures are primarily Hindu are future-oriented.

When establishing rapport, asking questions and listening to responses of interviewees from different cultures, the interviewer needs to be sensitive to how the perception of time plays a role in the responses. Time is also a critical component of social events but differs by culture. To show up late for an appointment in Latin America is a show of respect but is considered rude in Germany (Samovar & Porter, 1991).

Anyone who has spent time in Italy knows that time is irrelevant. The Italians gave the world the Renaissance, but they did it without timepieces. If businesses, museums, restaurants, trains, and the like operate on a schedule, it is difficult to discern. Shops and museums often close on a whim of the merchant or curator. In Germany, time is of the essence. Efficiency is prized. If a train is scheduled to leave at 3:10 P.M., for example, as soon as the second hand ticks to 3:10 P.M., the train pulls out. If a technician says he or she will be at a client's house for a service call at noon, that client's doorbell will ring at noon.

Time is a critical component of many types of interviews. When investigating crimes, whether it is one incident of sexual assault or multiple incidents of domestic abuse or child sexual abuse, time is critical in placing the offender and victim at the same location at the same time. If probation officers are discussing conditions of probation, time has an important role in completion of community service, finding a job, attending alcohol treatment sessions, and keeping office appointments.

It is not sufficient to expect individuals with different time perspectives to assimilate to the monochromic, futuristic approach of the United States. The interviewer must ask questions to clarify the interviewee's approach to time. Questions surrounding family events, birthdays, and time related to work will help clarify the interviewee's understanding about time as viewed in the United States, and also will help the interviewer understand how the interviewee's culture views time. As discussed in Chapter 3, it might be useful to use other events to anchor time, such as birthdays or holidays of that specific culture.

EXERCISE 8.2

Expressions of Time

Activity: Students brainstorm expressions using the word *time* or expressions relative to time.

Purpose: To facilitate students' appreciation about others' perception of time.

Divide students into small groups and tell them to come up with as many expressions as possible surrounding the word or concept of time such as, "What's happening?" "doing time," or "serving time."

The instructor should lead a discussion about what the different terms mean. Did the students come up with terms that are more polychromic or more monochromic? Did they think of "time is money," "in a little while," "fixin' to," or "ribbon of time?"

Nonverbal Communication Differences

As covered in Chapter 5, nonverbal communication is used to send and receive messages, and to make judgments and decisions concerning first impressions of others and perceptions of the experiences of others. The quality of relationships with others is judged through nonverbal messages. Emotions are reflected in posture, facial expressions, and eye contact. There is a strong link between use of nonverbal communication and culture. Nonverbal communication, like culture, permeates everything, involving all stimuli within a communication setting, generated by the communicators and their use of the environment.

Nonverbal communication is used for a variety of purposes. It repeats what was said verbally, clarifying and emphasizing. It complements and accents what was said, such as clapping hands while congratulating.

Of course, sometimes nonverbal communication contradicts what was said verbally, and the listener must decide which to believe. Faced with an individual, male or female, whose eyes are watering, lower lip quivering, clasping and unclasping his or her hands together, and saying, "No, I do not think your information affects me in any way," most people would be likely to believe the individual's nonverbal cues. The information bothers him or her a great deal.

Emblems are forms of nonverbal communication that substitute for the spoken word, nodding for agreement or waving a hand nonchalantly as if the interviewer is being dismissed, or the question is irrelevant in the eyes of the respondent. Nonverbal turn-taking signals, such as pausing and looking directly at the listener expectantly, provide a speech-regulating function.

Nonverbal signals often perform a linking function between movement and a communicator's emotional state, such as tapping or drumming fingers (Samovar & Porter, 1991). For example, what movie buff can forget that memorable scene from the move, *The Caine Mutiny*, where Humphrey Bogart, as Captain Queeg, is on the witness stand? Bogart has metal balls in his hands, and he keeps tumbling them—a dramatic portrayal of nervousness and deception.

The Kennedy-Nixon presidential debates were broadcast on television in black and white. Kennedy appeared calm and in control. His makeup was impeccable. Nixon, on the other hand, looked nervous and perspired freely. Was this the man to lead the nation? On election day, the voters said no. While many analysts said Nixon won the debate based on his spoken words, others countered that he lost the Presidency in that debate by his nonverbal cues. Folks who listened to the debate on radio overwhelmingly agreed that Nixon had won the debate. While those folks who watched it on television gave Kennedy a wide margin of victory (Schroeder, 2008).

In the play *Frost/Nixon*, playwright Peter Morgan as much as acknowledges the powerful role of nonverbal cues (Morgan, 2006). When former President Richard M. Nixon agrees to a series of interviews with the glib

Britain David Frost, one of the contractual agreements was to permit Nixon to have a handkerchief hidden under his leg. Nixon would be allowed to wipe the perspiration from his face, but those scenes would be edited out when the interviews aired on television.

Given the variety of differences in nonverbal cues based on culture, when interviewers are aware they will be interviewing an individual from a specific culture, at a minimum, they must remain diligent to the nonverbal cues they are projecting and remain sensitive to the possibility of differences.

To be an effective interviewer, the interviewee's culture and use of nonverbal cues should be studied and carefully observed within the context of the interview.

The following examples should not be considered comprehensive, but rather absorbed to increase appreciation for the diversity of nonverbal cues that are observed from all over the human body. Researchers in the area of linguistics and psychology have identified more than 700,000 separate physical signs (Edelmann, 1999).

Body Space   Space communicates intimacy by actions that simultaneously communicate warmth, closeness, and accessibility. Immediate behavior includes open body positions and allowance of closer distances between the speaker and listener. If accepted, individuals tend to reciprocate. Cultures that display considerable interpersonal closeness or immediacy are *high-contact cultures*. In cultures such as found in Latin America, South America, southern and eastern Europe and Arabic countries, individuals stand closer, frequently touch, and prefer sensory stimulation including odors.

In contrast, *low-contact cultures* are countries in Asia and northern Europe. Space around individuals is highly respected; touch is usually unacceptable. United States and other areas of Europe are generally high-contact, probably due to generational and internationalization influences.

Personal body space for United States citizens of about 18 inches is reserved for intimate conversation. Social space for casual interactions is 18 to 48 inches. Body space varies by culture and by social status.

Higher-status people assume and are granted more personal space than people of lower status. Individualistic cultures demand more space than such collectivistic cultures as Latin America and Israel. Particularly in collectivistic cultures, space can be equated with respect. For example, teachers or superiors are given more space than students or line personnel.

In other words, the interviewer must pay attention to body space perceptions. Sitting too closely may make the interviewee uncomfortable, but sitting too far away may be interpreted as aloofness. Before interpreting certain nonverbal cues as signs of anxiety, the interviewers need to be certain the chairs are not seated closer than is comfortable for the interviewee. Could there be cultural taboos being violated that is causing cues of anxiety?

When an interviewer is preparing for an interview, spacing of the chairs is important to convey level of comfort or tension the interviewer wishes to convey. The cultures of some Middle Eastern countries will allow and expect to sit close enough to observe the dilation and constriction of the interviewee's pupil. Tension is not likely to occur with interviewees from Middle Eastern cultures when the interviewer pulls the chairs close together (unless one of the communicators is a female).

Bowing in Asian cultures is different based on rank. Lower rank individuals begin the bow, and their bows are deeper.

Status is shown in the United States by who goes through the door first and who sits down first. Status also is determined by proximity to the "star" in the group. The closer a person sits to the chief executive officer (CEO), for example, the more influence that person has. The same generalization goes for offices. The closer to the boss, the more likely that the individual has more influence in that business than someone in a different part of the building.

In the Middle East, status is indicated by which individual views one's back when one turns around. When acceptable, the higher ranking individual is likely to show his or her back. Similarly in Asian cultures, backing out of a room is expected from one who is lower rank. Squatting, which is rarely used by U.S. citizens, but is common in Mexico, is used by the U.S. Border Patrol to detect illegal Mexican aliens (Samovar & Porter, 1991). The squatting individual is likely to be an illegal immigrant, who is not aware of U.S. mores.

In some cultures, it is rude to show a person the bottom of feet or shoes. For example, in late 2008, during a U.S. press briefing in Baghdad, an obscure Iraqi television journalist, 30-year-old Muntathar al-Zaidi, hurled his shoes across a crowded room at President George W. Bush. The Iraqi journalist was expressing his extreme disdain for Bush and his policies and his frustration at the war in Iraq. In Arabic society, merely showing a person the bottom of a foot is considered an insult. Middle Eastern cultures consider the shoe and the bottom of the feet to be filthy because they touch the ground and all the litter strewn on the ground. Muntathar al-Zaidi was arrested on December 14, 2008, charged with assaulting a foreign head of state. He was sentenced to a 3-year prison term. al-Zaidi's sentence was reduced to 9 months because he had no criminal record (Keyser, 2009).

As described in Chapter 4, the first 4 minutes of an interview are critical. The interviewer and interviewee are observing each other. It is similar to the beginning of a heavyweight boxing match. The contestants begin slowly, just feeling each other out. The interviewer must develop rapport and credibility within those few minutes. Understanding the nonverbal communication surrounding greeting and introductions may enhance or extinguish opportunities for an effective interview.

Facial Expressions Some emotions, such as anger, fear, sadness, agony, surprise, disgust, and enjoyment, are universal; however, individuals from different cultures vary in the inferences they draw from emotional facial expressions. One cultural group may show negative emotions to a situation that another cultural group views as positive. Response time, range, and meaning of expression may vary.

There are cultural differences in unmasking emotions. Mediterranean cultures readily express signs of grief and sadness; men will cry in public. In the United States, men, especially, suppress a show of emotion, especially sorrow, anger, or disgust.

Asian cultures are taught to conceal, or save face, in public. They keep their facial expressions neutral in public (Samovar & Porter, 1991). In countries in which power, prestige, and wealth are unequally distributed, oppressed people become skilled in decoding nonverbal behavior. Lower level individuals are expected to show only positive emotions toward high-status people. The continuous smiles of many Asians are efforts to produce smooth social relations.

Facial expressions, as with other nonverbal communications, depend on the setting: business, home, informal, and intimate.

Eye Contact   Messages sent with eyes are infinite, communicating degrees of attentiveness, interest, and arousal; influencing attitude change and persuasion; regulating interactions; and communicating emotions (Eakins & Eakins, 1978, as cited in Samovar & Porter, 1991).

Direct eye contact varies among cultures. Westerners expect people to make direct eye contact, while Japanese, Chinese, Indonesians, and rural Mexicans avoid sustained and direct eye contact. Arabs look directly and for sustained periods of time into the eyes of the person with whom they are communicating.

Different cultures use eye contact to represent power. In the United States, downcast eyes and lower body positions are expected by individuals in subordinate roles. In Japan, downcast eyes are seen as a sign of attentiveness and agreement. Lower level body position signals acceptance and respect and may be perceived as a sign that a person is trustworthy and accepting (Gudykunst, 2003).

Within the United States, there are cultural differences. Many Native Americans believe that direct eye contact is offensive, and consequently, they avoid staring. For the Navajos, staring is considered an "evil eye" and implies an aggressive assault. African Americans use continuous eye contact when speaking but less when listening, while Caucasians use more eye contact when listening and shun continuous eye contact when speaking. These differences contribute to misunderstandings, particularly between African American and Caucasian communicators. African American speakers' less eye contact when listening is perceived by Caucasians as disinterest. The longer eye contact by Caucasians is interpreted as hostile by African Americans.

These variations emphasize the importance of change in interviewee's facial expressions and eye contact. The interviewer cannot relate deception to specific facial expressions or the interviewee's lack of eye contact. There is no definable set of nonverbal cues that signals deception or anxiety (Virj, Edward, Roberts, & Bull, 2000). Taking time to develop rapport will help the interviewer observe the interviewee. How much eye contact is used when discussing the individual's family or job? What facial expressions does the interviewee use to show grief or interest? Did the nonverbal cues change? These are baselines that will help the interviewer become sensitive to changes in the interviewee's expressions.

Effective Intercultural Communication

Effective intercultural communication needs two ingredients: knowledge about the interviewee's culture and the ability to clarify meaning and perceptions. If these two ingredients are placed on a graph, it is possible to see how the effectiveness of an intercultural interview is on a continuum (**Figure 8.1**).

The most effective communicator should have acquired a great deal of knowledge about the interviewee's culture, values, traditions, customs, and practices through reading information about the culture and talking to experts and ordinary people from the culture. New interviewers may only be able to read general information about the culture and perhaps talk to another professional who works with individuals from that culture. Effective interviewers who know they will continue to interview people with certain cultural backgrounds will continue to develop knowledge, exploring the distinctions among regions, and even learning the language.

While it is important to seek clarification through paraphrasing, probing, and summarizing with interviews within the same culture, the ability becomes even more critical with intercultural interviews. The interviewer has

Figure 8.1 Effective Communicator

High Knowledge

(Informed, insensitive) | (Informed, empathic)

Low Ability -- **High Ability**

(Uninformed, insensitive) | (Uninformed, empathic)

Low Knowledge

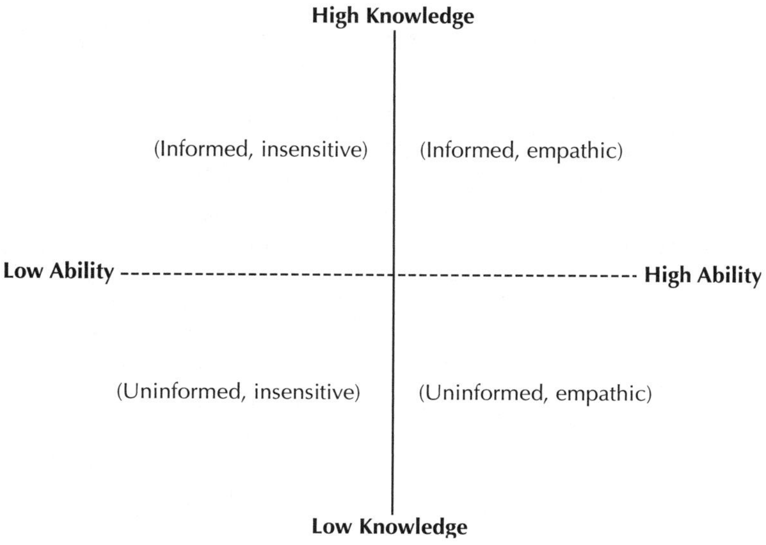

to remain diligent to the probability that nonverbal cues should be interpreted differently; the interviewee may interpret the questions differently from the meaning of the interviewer; the interviewee's response may be based on misinterpretations of the question; and certain words, phrases, and gestures may be sensitive or even taboo.

Humor is a good example. Different cultures find different things funny, and what some cultures consider funny, others find offensive. In general, humor should be avoided.

Developing Rapport: A Good Place to Start

Acknowledging the lack of knowledge about the interviewee's culture but showing a desire to learn will help the rapport-building process and also help the interviewer avoid cultural "landmines."

The introductions should begin with asking the interviewee how he or she wants to be addressed. Is the interviewee more comfortable called by his or her first or last name? What title, if any, do they prefer in front of the name? This question sends an immediate message of openness and desire for information to the interviewee. It shows respect for the interviewee and facilitates the development of an environment that the interviewee will feel safe enough in to give valid and complete information (Yeschke, 1997).

The interviewer's own introduction should include information that will briefly state credentials and title. Interviewees, especially of different cultures, also will want to know how the interviewer should be addressed. In cultures in which rank is important, the interviewer should either use appropriate rank and title, or specifically tell the interviewee how to address him or her: "My name is Detective Pauline Brandon. Call me what you are comfortable with—Detective Brandon or Pauline" (if addressing a victim).

Another introductory statement might include an invitation for the interviewee to be sure to voice any concern he or she might have about what is asked or how it is asked.

General Hints for Effective Intercultural Communication

The interviewer should speak slowly and articulate carefully, using simple vocabulary. In many cases, interviewees will have limited English skills and must mentally translate through their own language.

The interviewer should avoid technical terms, slang, or idioms. If specialized vocabulary must be used, such as legal terms, they should be explained, allowing questions for clarification.

Sometimes the questions may appear unnecessary and redundant, but it is important to be patient. For individuals who are not fluent in English and the local culture, it will not appear as simple and straightforward.

Many people from other cultures may be reluctant to ask questions and may give the impression that they understand. It might be useful to have handouts to supplement oral instructions or presentation of information.

People from other cultures often respond in the affirmative, but agreement does not necessarily mean understanding or acceptance. In some cultures, it is discourteous to disagree. Asking them to summarize what has been discussed is a constructive way to assess their understanding.

As has been discussed earlier in this chapter and in Chapter 4, detecting deception from behavioral cues is not a simple task. Interviewing across cultures with various age and gender types makes the task more difficult. The interviewer should not make any conclusions about deception based on nonverbal cues under any circumstance, but rather probe areas in which the interviewee's nonverbal cues appear to change. Using the baseline that is developed during rapport building, the interviewer can watch for changes from baseline behavior. These changes only suggest that the interviewee has apprehension or anxiety, but do not necessarily mean deception. The interviewer needs to consider other factors that could contribute to the anxiety and attempt to decrease those factors.

Space between the interviewee and interviewer could be the cause of anxiety. Beginning an interview about 6 feet apart should not frighten or anger an interviewee no matter the culture. As the interview progresses, the interviewer can consider moving closer to convey warmth and to maintain focus on the interview. Careful attention to the interviewee's response if the space is narrowed is paramount (Yeschke, 1997).

While it is important to be careful about generalizations of different cultures, as mentioned earlier, any awareness of some general facts about a culture will begin the process of understanding and help keep the interviewer sensitive. Given emerging differences among generations, it is difficult to be completely interculturally savvy. **Table 8.1** displays general characteristics of some global cultures.

EXERCISE 8.3

Origin of Cultural Beliefs

Activity: Brainstorm with students' different stereotypes and how they are formed.

Purpose: To promote increased student awareness of their stereotypes.

   Ask students to brainstorm adjectives/descriptors that other cultures have of U.S. citizens.

   Ask students to brainstorm adjectives/descriptors that U.S. citizens have of another culture (select one: Mexican, Iraqi).

   Ask students to brainstorm how others have created their stereotypes, e.g., movies, TV.

   Ask students to brainstorm how we have created our stereotype.

Table 8.1 General Characteristics of Global Cultures

Global Cultures	General Characteristics
Asian Cultures	
Japanese	1. Strong identity is given to the group. 2. Personal relationships are important. 3. Age and tradition are honored, but not at the expense of progress. 4. Hard work and devotion are strong motivators. 5. Social- and self-control disguise highly emotional situations. 6. Laughter can mean happiness but is also a sign of embarrassment. 7. It is impolite to yawn in formal situations.
Chinese	1. Betterment of the group is emphasized over the individual. 2. Practical business orientation with a sense of fairness is important. 3. Emphasis is placed on proper etiquette with sensitivity to customs, even temperament, dignity, reserve, patience, and persistence. Honor and face is paramount. 4. The family is important with respect for elders and those in authority. 5. The family name and then given name is stated. 6. The open hand is used for pointing rather than the index finger. 7. Greetings often are head nodding with handshaking less frequent.
Korean	1. Traditional respect for elders and those in higher status is paramount. There are different vocabulary and verb forms for superiors. Conversation usually is not initiated toward elders, and eye contact is avoided. The elders are referred by their titles such as teacher, rather than their names. Rather than possibly insulting the elder by stating, "I don't understand," the younger person will nod politely and remain silent. 2. Personal opinion will rarely be expressed so as to not sound presumptuous except in areas of justice, integrity, and morality. 3. Open disagreement is avoided.
Vietnamese	1. Family comes before self-interest; however, respect for self as well as others is important. Courtesy toward others is an important point of etiquette. 2. Looking somebody directly in the eye is considered disrespectful, especially from a younger person. 3. Problems or the expression of feelings are not easily disclosed. Antisocial emotions such as anger and hostility are to be concealed. A smile can mean stoic behavior toward adversity or may conceal hostile or angry impulses. Unresolved conflicts are turned inward and sometimes result in suicide. 4. Modesty and self-deprecating behavior is shown toward personal accomplishments. 5. Females are to be modest, and males are traditionally dominant over females; however, in the United States there is tension between the sexes because males have had to accept lower-class jobs. 6. Social touching is limited. Even shaking hands with the opposite sex can be uncomfortable. Same-sex touching, such as holding hands or walking arm in arm, is common.

Table 8.1 General Characteristics of Global Cultures (*Continued*)

Global Cultures	General Characteristics
Indian Culture	1. Family oriented with strong respect for age is central. Parents have a strong influence over children. 2. Titles such as doctor or teacher are used to indicate respect. 3. Strong beliefs in simple material comfort and rich, spiritual, and philosophical accomplishments are traditional. 4. Communication is candid, but not verbose. Women are more subdued, and men more gregarious. 5. There is an expectation of working hard, but work is not to be stress-oriented.
South American (with variations by country and culture)	1. The extended family acts as a support group. 2. Personal issues are mixed with professional ones, and it is acceptable to be sociable in business settings. 3. Criticism and pushiness are offensive, and most settings are relaxed. 4. Social touching is acceptable, especially of same sex. Conversations are conducted in close proximity. 5. Friends and relatives are greeted with a kiss.
Arabic Culture	1. Language is important with flattery and compliments used extensively. 2. Repetition is important such that invitations and refusals must be repeated to be considered real. 3. Criticism is only appropriate in private. Bluntness is disrespectful. 4. There is fascination with foreigners and their language. 5. Any expression of admiration for a possession is interpreted to mean that the possession should be given to the admirer. 6. The level of emotion is high for either ecstasy or depression. 7. Social touching of the same sex is common, and close proximity is used when communicating.

Conclusion

Communicating between different cultures is challenging. The interviewer must be educated in the culture or be sensitive to the differences in the cultures.

In criminal justice interviews, accurate and complete information is crucial and this places a higher level of importance on two-way communication and understanding. The interviewer must learn to become highly attuned to differences between cultures and acquire the ability to communicate questions in an understandable fashion. The ability to promote comfort in interviewees so they will ask for clarification and other questions is critical for effective interviewing.

EXERCISE 8.4

Observations

Activity: Observe interactions with different cultures.

Purpose: To provide a real environment in which the students can compare how individuals from different cultures interact.

Suggest that students attend their favorite ethnic restaurant (it cannot be a fast food drive-through) and observe the couples and families who also are having a meal. How do the men and women interact? Observe their eye contact, facial expression, body space, and clothing. How do they interact with their children? How do they allow the children to interact with them? How do they interact with the servers?

ALLEN'S WORLD

It is difficult to relate much of this chapter to the world of interviewing for a private investigator.

Yes, I deal with cultural differences, and yes, I have to account for that, at times, in my interviews. Often the differences I find are regional and socioeconomic.

I never correct grammar or pronunciation. My notes, or tape recordings, reflect exactly what is said and how it is said. I often hear the word *axed*, as in, "Can I axed you a question?" when I know the person means *ask*. I never change that word.

Do not come across as condescending. Condescension kills interviews.

Do not patronize. Patronizing kills interviews.

If I need to interview an African-American woman, and her man is present, I must make a choice. I can diplomatically ask the woman if there is someplace private we can talk, or I can advise the man that he can stay for the interview, but under no circumstances is he allowed to prompt the witness or go so far as to answer questions for her.

If the man and woman both have knowledge of the incident I am investigating, I always separate the man and the woman and conduct independent interviews. I just explain my reasons.

Much of the above applies to my need to interview a Hispanic man or a woman. If either speaks English, no matter how broken, I do not correct them—notes, tapes, verbatim—as long as I am confident they understand the question, and I understand the answer. It helps that I speak a little Spanish. Yo hablo español (I speak Spanish) un poquito (a little). Como esta usted? (How are you?)

I have, on occasion, had to interview folks who knew no English. I always hired an interpreter.

One time, while writing an investigative article on the Praise the Lord (PTL) ministry for the *Charlotte Observer*, I had to do a telephone interview with a Korean minister. I called the local university, the University of North Carolina at Charlotte, and found a Korean instructor who spoke English. That interpreter came to the *Observer* at midnight, we called Korea and taped the interview.

I often show my state-issued credentials before the interview. As Dr. Lord suggests, at this stage I let the subject address me any way he or she chooses—whatever makes that person comfortable. I am not easily offended.

If the incident I am working on involves a sexual allegation, I must find a way to make the woman feel comfortable. Men will discuss sexual charges easily and graphically. For women, it is harder, especially when the interviewer is a male. I do not have specific recommendations here. Much depends on the age of the woman, the surroundings of the interview, and how long it has been since the incident took place.

Never suggest to the woman you do not believe her, regardless of what she says. Most women feel victimized by the sexual offense. If she suspects she is not believed, it creates a second feeling of victimization. An Asian woman may have difficulty talking about sexual allegations. She may need the man in her life to assist her.

Interviewing genteel southern women creates a unique problem. In general, I have found that southern women do not want to appear hostile or assertive, and yet they are reluctant to provide information.

One time, I needed the picture of a heart attack victim. He fell down at the Central Y (now called the Dowd Y), and the staff saved his life. This was going to be a great human interest article.

I called his wife and asked if she had a picture of her husband. She told me she would look for a picture.

I called back. She was still looking.

I called back. She was still looking. We were getting closer to deadline.

I called back. Her adult daughter answered the phone. I told her I needed a picture of her father, and she said she had one I could have. I sent a taxi to the house, and the driver came to the *Observer* with the picture.

The next day, the mother called the editor of the *Observer* and voiced a complaint. She requested a meeting.

"I told Allen I didn't want him to have a picture of my husband," she told the editor. "So, he snuck around my back and got one from my daughter. I want it back."

The editor asked me if this was an accurate representation.

"Ma'am," I asked, "when exactly did you tell me you didn't want me to have a picture of your husband?"

"I told you three times I couldn't find his picture," she said.

I was dumbfounded. Telling me she could not find the picture was the equivalent, to her, of telling me she did not want me to have a picture.

The editor politely told the woman he understood her concern and told her he would deal with me.

She left. The editor told me I had done a good job.

Southern women often talk indirectly. Be careful during these interviews.

I almost always dress casually. I wear a University of Florida shirt regardless of whom I am interviewing. In the south, a Gator shirt almost always starts a conversation. A lot of folks matriculated at southeastern conference schools. The Gator shirt gets them going. People like us or hate us. It doesn't matter to me, as long as this shirt becomes an ice breaker for me.

One time, I was solicited to come to the office of a CEO whose company was a victim of employee theft. He said I was one of two private investigators being interviewed to find the culprit.

Well, when I walked into his office, the room was a dazzling orange. He was a graduate of the University of Tennessee, nicknamed the Volunteers, arch rivals of the Gators. He had Tennessee paraphernalia everywhere.

I took a look around and said to him, "I'm not going to get this job, am I?"

What happened is interesting. I walked around his plant and quickly saw how easy it was for an employee to steal merchandise. I told him voluntarily how to lock the barn door so no more horses got away. He took my recommendations. I lost the job by giving away information for free.

If the setting is not one of my choosing, I will look around and find something to start the conversation: pictures of the interviewee with politicians, sports paraphernalia, family photos, or deer antlers on the wall. If I find something I do not know what it is, I ask. People love to talk about themselves, and this gives them a way to feel important and for me to disarm them.

I place more emphasis on nonverbals than Dr. Lord does. Over the years, I have developed my own internal lie detector. These nonverbal cues often develop into deceptive statements.

I do not trust people who smoke pipes, wear pinkie rings, or drive Cadillacs. These nonverbals suggest a level of vanity that makes me uncomfortable—and suspicious.

I am wary of people who

- wear toupees;
- do not look at me during the interview;
- have a lot of pictures in their office of them with famous people;
- have a lot of trophies in their office;
- sit behind a desk that is elevated while my chair is on the floor a little lower than the desk (this is a power play, and it makes me nervous); or
- have a window behind their desk, and consequently, the sun comes in and is shining in my face.

I cannot strap my respondents into polygraph machines. I cannot do voice stress analysis.

What I can do is watch for nonverbal cues, listen carefully to what is being said and how it is being said, and then trust my instincts.

I once asked Abe Rosenthal, who at the time was the managing editor of the *New York Times*, how he made tough decisions about printing articles and classified information and revealing sources.

Rosenthal told me he relied on the "stomach test." Nobody ever told me about that test— not in journalism school, not an editor, not any attorney I ever worked for.

"What's the stomach test?" I asked.

"If my stomach hurts, I don't do it," Rosenthal said. "If my stomach doesn't hurt, I go for it."

I think that was Rosenthal's way of telling me, trust your instincts. And I have, when I walk out of an interview, and my stomach is churning. That happens too often.

Interviewing for Defense Attorneys

I am Allen Cowan, the coauthor of this book, a state-licensed private investigator in Charlotte, North Carolina. I have been working for criminal defense attorneys for more than 20 years. This chapter focuses primarily on the experiences of defense attorneys, building upon the "Allen's World" sections found throughout this book, which are based on my experiences as a private investigator and investigative reporter.

Keep in mind that as a private investigator, I am not bound to honor guidelines established by police departments, the court system, and the U.S. Constitution. For example, I don't have to read the *Miranda* warnings to someone he wants to interview.

I cannot break the law, but I can work in the grey areas where hair splitting becomes an art.

Throughout this book, in most instances, pseudonyms have been used to protect sources, clients, defendants and others who may have been useful during my career.

Introduction

Sam Spade
Philip Marlowe
Dick Tracy
Paul Drake
Magnum PI
Sherlock Holmes
Hercule Poirot
Ellery Queen

Fictional investigators, all of them. Readers should etch one tenet into their minds when thinking about private investigators (PI). And that mantra is: little of what students read in adventure stories, see on television or in the movies, or hear on the radio is believable.

Rarely are there last-minute revelations or the discovery of vital clues that identify the perpetrator or confessions from the witness stand.

Investigators do not solve cases in minutes as mandated by the length of a television script or in 350 pages, the average length of most mystery-fiction novels. The defense attorneys we work for do not expect us to complete cases in such abbreviated time frames. It is grind it out, grunt work, creating a mosaic of an event for the defense that might create reasonable doubt if the case goes to trial.

Many PIs do not own guns. Most of us do not smoke, drink, drive fancy cars, or, sad to admit, have gorgeous women meeting us in bars for an evening's entertainment.

It just does not happen.

Despite the glamorous lifestyle represented in the fictional world, some people consider what we do as tedious and boring. Not me. I love what I do, and at times, I feel like a crook for sending out bills for service rendered.

In this chapter, I'll discuss why defense attorneys hire PIs, and what one district attorney thinks about the use of PIs and the evidence developed by them. The objective of PI interviews is usually to elicit relevant details that will support information from the client based on what witnesses know personally and to develop leads to other witnesses.

I will include simple definitions of terms I think students might not be familiar with, emphasizing that I am not an attorney, and the meanings and nuances of legal terms can be debated among intellectually honest advocates with different agendas to advocate. Consider the U.S. Supreme Court. If legal definitions were so clear cut, we would not need the Court.

Use of Private Investigators by Criminal Defense Attorneys

One of the criminal defense attorneys I work with, I'll call him Tony, spent nearly 5 years as an assistant district attorney (ADA) in Mecklenburg County before shifting to the more lucrative private field. He is now a partner with a law firm that specializes in criminal defense. He and I work on, what Dr. Lord calls, "the dark side."

Tony is one of the few attorneys in Charlotte who recognizes the value of private investigators.

Why does he use a PI?

"I have to know the facts of the case," Tony explains. "I can't always rely on information given to me by the other side (long pause) or by my client."

How does Tony decide when he needs a PI?

"When the facts are in dispute that a PI might be able to clarify—and if the client has the money for a PI."

"In cases where identification is an issue, if the victim is a stranger, then I like a PI to talk to the victim who has identified my client as the perpetrator. The quality of the identification is always a question."

Why does Tony not do the investigation himself?

"I have no training. I have no expertise in figuring stuff out. I don't know how to put the information together. . . . . . I don't have the time. . . . . . . I can't put myself in the position of becoming a witness."

How much faith does Tony have in his private investigators?

"As a general matter, I'll trust my investigator's interview more than I'll trust the police interview. The police often don't conduct balanced interviews or write fair reports."

Does Tony tell his investigators what to do?

"Sometimes. Where there is a point of legal consequence, then I tend to give more feedback and direction."

For example, Tony had a case involving the use of a stolen credit card. The police conducted a photo lineup with the clerk at the store where the card was fraudulently used. That photo lineup violated every protocol for proper identifications. The witness identified Tony's client as the offender. Tony sent me to see the same witness to conduct another photo lineup, using the same pictures. In this instance, Tony gave the investigator (me) specific instructions on the proper way to conduct a photo lineup so he could contrast before a jury what the police did with what the PI did. In this instance with the PI, the witness did not identify Tony's client.

One District Attorney's Perception of Private Investigators

Peter Gilchrist III has been Mecklenburg County (Charlotte) North Carolina District Attorney since 1975. On December 4, 2009, he announced that he would not seek another term. An elected official, he often ran unopposed. His office is divided into divisions specializing in areas including prosecuting murder suspects, rape suspects, drug offense suspects, and property crime offenders.

I began my interview with Peter Gilchrist, as Dr. Lord often suggests, with an open-ended question.

"What do you think of private investigators?"

"I think they don't have a good reputation."

In general, Gilchrist says that since defense attorneys have their own agenda, he might listen to their argument, look at their material, and decide if it has credence.

"I think some lawyers tell private investigators what they want and then the private investigator goes and finds it," Gilchrist says.

When will Gilchrist listen to a defense attorney?

"It depends on the case. It depends on the lawyer. It depends on who the private investigator is. How thoroughly was the case investigated? I decide on a case-by-case basis."

Gilchrist knows there are attorneys, many of them former assistants who he trained and who worked for him before going over to the dark side, whose integrity is beyond reproach. If one of these attorneys calls, Gilchrist is more apt to listen and re-evaluate a case.

It also depends a great deal on who investigated for the police.

"Who was the officer?" Gilchrist says. "And how good do I feel about that?"

Gilchrist knows—as do most of my colleagues—that there are attorneys who manipulate information so it presents their client in the best possible light. So he might listen to a presentation and reserve the right as to how to handle it.

Gilchrist knows that there are PIs—as do most of my colleagues—who only turn over favorable information to the attorney of record and withhold information that might be useful to the prosecution. It all goes back to ethics, our own internal, moral compasses.

One attorney told me a story of three men charged with raping a woman. Each defendant had a separate defense attorney, but the three attorneys worked together.

First item on the agenda, they called in a PI. All the investigative work was done by a Charlotte-based PI, a great investigator, but with a reputation for cutting corners and crossing the line. This PI came up with unimpeachable evidence that the three men were innocent. The woman was lying—no doubt about it.

The attorneys, however, knew this PI was not in good standing with Gilchrist's office. So, they created a PowerPoint presentation including the written documentation and statements. They turned all of this information over to an ADA with one exception. They intentionally did not tell the ADA the name of the PI who did the investigation. They knew with certainty that if the PI's name was revealed, the new evidence would be considered tainted and might not even be examined.

In this instance, the ADA looked at the new evidence and dismissed all of the charges against the three defendants.

The Role of the PI for Defense Attorneys

Jodi McMaster states that litigators commonly note "a case never looks as good as the day it walks in the door" (2006, p. 37). I do not agree with her. Using the proper and dogged interviewing techniques detailed in this book, students should learn, especially in cases I discuss that ended in dismissals, that many of my cases look much better after the day it walks in the door.

Having made the proper disclaimers—repeating that I am not an attorney and none of what I say is intended as legal advice—let me tell you about some of my most interesting cases in a career that began in March 1989; cases that will educate, create moral dilemmas, hopefully entertain, and probe the gray areas of the criminal justice system.

I will skip my journalism career, which began in 1968 after graduation from the University of Florida. Suffice it to say that those newspaper years gave me the best foundation a private investigator could have. I learned how to track documents, use my sources, interview, turn over rocks, and follow the cracks, not always knowing where those cracks might lead.

When I could promise confidentiality, and the individual discovered that I was willing to protect him or her, even the most reluctant witness became an unstoppable torrent of information.

What is the promise of confidentiality? I tell a source that under no circumstance will I reveal who gave me information. Journalists have gone to jail rather than reveal sources. There are instances as a private investigator that I can tell a witness, "Nobody will know where this information came from. So why not help me out?"

I spent a year studying law at the University of Michigan in Ann Arbor while on a fellowship as an investigative reporter. That legal education allows me to meet with attorneys, understand what they need for trial, know how to get that information, and ensure it will stand up to rigorous cross-examination by any prosecutor.

My work load generally falls into three categories: domestic, personal injury, and criminal defense. All three require excellent and creative interviewing skills.

Domestic   All domestic cases are handled in civil court. Two people dispute alimony, child support or custody, or the distribution of marital assets.

For me, these cases usually begin with a phone call, either from an attorney or directly from a client who does not have an attorney. A husband or wife thinks a spouse is cheating, or a boyfriend or girlfriend thinks his or her significant other is cheating. If the call is from a client, I almost always implore that person to talk to an attorney before retaining me. Most often, what they ask me to do has little or no legal value. I do not want to waste their money and have them angry with me later because my work product turns out to have no legal value.

I arrange a meeting and begin to prepare questions for the information I will need.

EXERCISE 9.1

Preparing for a Domestic Meeting

Activity: The students incorporate what they have learned into initial meetings with clients of an alleged infidelity situation.

Purpose: To apply basic interview introduction techniques.

The students are divided into groups, or the hypothetical situation can be part of a group discussion.

A husband or wife thinks a spouse is cheating and has phoned for help.

1. As a PI, what would you want to know before the phone call is completed?
2. If you decide to meet, what would be the objectives of the meeting? What questions need to be prepared in advance?

The following is one of my cases. The names have been changed.

In early 1990, a prominent Charlotte attorney called me in to work on a case.

When I entered the conference room, seated next to the attorney was a drop-dead gorgeous woman—perfectly made up, high cheek bones, elegantly dressed. She oozed breeding and wealth. Let us call her Nancy for the purpose of confidentiality. Nancy was married to Oscar. Oscar was a prominent businessman in Charlotte. Business community folks would immediately know who he was.

Oscar was rich. He was a partner in a lucrative business. The custody of two children, alimony, and child support depended on our investigation. North Carolina is now an equitable distribution of assets state, where assets obtained during a marriage, in general, are shared equitably upon divorce. At the time of this case, that was not state law, and divorcing spouses could claim bigger shares of the pie.

Nancy felt her husband was cheating. She wanted him followed. Money was not an object.

"Nancy, what makes you suspect Oscar?"

Nancy told me that her husband had tickets to the Charlotte Hornets NBA games, but that he would often come home and not know any details.

"How often does this happen?" I asked. "Is there a pattern?"

"Not often," Nancy said, "and there is no pattern. He'll just call at the last minute and say he decided to go to the game."

"Anything else?" I asked.

"He sometimes calls and says he will be working late," she said. "If I call at work, whoever answers says he's meeting with a client. He calls back a few minutes later, but it seems as if the call is from a cell phone and not from his office. I don't believe he is at the office working late."

Nancy brought a picture of her husband. I always ask for a physical description because a picture does not tell me how tall a person is, how much he weighs, or any details that may have changed since the picture was taken.

I asked standard questions: Where does he work? What make and model of car does he drive? What are the numbers on his license plate? Does he have a real job? In other words, does he have to be at a fixed location during the day, or does he have the freedom to roam the city?

I gave my standard spiel. We would do a domestic surveillance. I partner with another fully licensed PI and not a trainee. I explained the costs, what information Nancy could expect, and that if necessary, my partner and I would testify if the case goes to court.

Nancy and the attorney gave me complete freedom to do what I felt was necessary. They did not limit my expenses or fees; there was simply too much money at stake.

I called my partner, and we decided to do lunchtime surveillance.

Surveillances are tricky. We might get caught, or "burned," by Oscar. I use another PI so we can trade off, and neither one of us will be behind the suspect for long; harder for us to get burned.

In some jurisdictions, I could possibly be charged with stalking. If we are doing our job properly, the subject will not know he or she is being followed. So I do not consider the possibility of being charged as a stalker.

I have to be clear what a public place is and where we are allowed to go and not go to watch the suspected spouse. The suspected spouse has some rights to privacy. What expectations do people have that what they are doing at any given moment is a private experience? What is a public place? If I can see the cheater and his lover well enough to take pictures and videos, I consider that a public place. His or her expectation of privacy differs from mine. And yes, I take as many pictures and/or videos as I can.

I want to be sure that my actions, including pictures or videos I take, will be admissible in court. *Admissible* is a term for evidence such as physical evidence, statements, and testimony that a judge will allow a jury to hear or see. A governing principle of admissibility is relevancy. To judge relevancy, admissible evidence must increase or decrease the probability of the fact or action in question. Although relevant, evidence also can be thrown out if the danger of prejudicing, confusing, or misleading the jury outweighs its persuasive value (McMaster, 2006).

I do not worry about admissibility in domestic cases. If we catch the person cheating, almost always these cases are settled mutually without benefit of court proceedings. The information I provide the attorney is used as leverage or to create pressure on the transgressor to become more cooperative.

I do not sit in on those conferences, but I can imagine the conversation. My attorney says to the cheater's attorney, "We can settle this privately, or we can go to court and show this video to a jury. It's your call."

As of this writing, my partner and I have never gone to court to testify in the instances when we have caught a spouse cheating.

My partner and I set up surveillance on Oscar at his work place. We both could see his car. One of us was on each side of the street, so regardless of which direction Oscar went as he left his office, one of us could quickly pick him up. We used two-way CB radios to stay in touch. Both radios were tweaked to maximize range and efficiency.

Oscar came out at lunchtime, pulled onto Independence Boulevard, and drove west. It appeared he might be heading for lunch. Oscar pulled off Independence into a shopping mall and drove his car to a remote corner of the parking lot. It was clear; this was not a lunch date. Oscar got out of the car and appeared to be waiting for someone. Within minutes, another car drove up, the driver's side door opened, and a woman stepped out. It was not Nancy. The woman went up to Oscar, threw her arms around him, and gave him a big hug and a kiss. She stepped back, and they appeared to be talking.

It did not matter, but we felt certain Oscar was in a public place. He had no right to privacy. We photographed Oscar and the woman together. We photographed his car and her car and their license plates for positive identification. Oscar got into her car and drove away. The woman drove away in Oscar's car.

In North Carolina, you need two factors of evidence to prove adultery: *inclination* and *opportunity*. The simple definition of inclination is a public display of affection. So, we had the rendezvous and a public display of affection—the inclination needed. Oscar could argue all he wanted that this was a business meeting, but the hug and kiss would belie any later protests.

We did not follow Oscar or the woman from the parking lot. It did not matter where they went at that point, and the less surveillance work we did would minimize the chances of being caught.

Now that we had her license plate, the next step was easy. I called the department of motor vehicles (DMV) in Raleigh, gave the associate the license plate number of the woman's car, and within seconds I knew her name. More importantly, I knew where she lived. In North Carolina, it is legal to obtain DMV information. I do not have any special privileges to get that vehicle information. I have an account with the DMV, as do most private investigators. Any citizen can get the same information, although it is a more cumbersome process. The citizen has to write DMV, telling them what he or she wants. DMV replies with a letter stating the fee involved. The citizen then has to prepay the fee, not knowing if he or she will get a hit or not. DMV responds with what information it has. This process can take weeks. Because most people are in a hurry to get DMV information, they go through a private investigator. Each state has a different system. Some do not consider motor vehicle information a public record and will not supply the information.

We now needed the second factor: opportunity. Again, a simple definition of opportunity is two people who are in a private place long enough for a sexual act to have taken place. Oscar soon provided evidence for that factor. Nancy called a few days later and said Oscar told her he was going to the NBA game. We did not follow Oscar from work, which, again, minimized our chance of being burned, or having the person we are following figure out he or she is being followed. Instead we set up a surveillance on the woman's apartment that was on the second floor of a high rise. We could see inside her apartment, and we could see the parking lot. Her car, the one she had left with Oscar, was in the parking lot. At a minimum, that was confirmation to which we could testify that they had to have met after we saw them together. How else could they have swapped cars?

After a short wait, Oscar drove up. The woman we had seen him with in the parking lot came out of the building and got into Oscar's car. We followed them into a restaurant. My partner and I went inside; we got a table with a clear, unobstructed view of Oscar and his date; and we ordered a meal.

Could we be charged with stalking? Oscar was so involved with the woman, he did not notice us. Also, he was in a public place, and we were not there to do any physical harm or to accost them in any way.

Oscar and the woman were sitting side by side. He had his arm around her shoulders. She had her right hand on his left thigh. They were joking, drinking beer, and eating. We made notes of all that we saw. Oscar and the woman eventually left. In cases such as this, I prepay for the meal and leave a tip in case we have to jump up and rush out. We did not follow Oscar and his date. We sped back to our vantage point at her complex, set up our video cameras with zoom lenses, and waited. I turned on the radio to find out the progress of the NBA game. It was early in the third quarter. Oscar only had about 90 minutes left before his wife would be expecting him home.

Oscar and the woman drove into the parking lot. Both got out of the car and went into the building. A few seconds later, a light came on in her apartment. We could clearly see Oscar and the woman. She went into what appeared to be a bedroom and began disrobing. Her blinds were open. Her lights were on. Oscar was still in what I would call the living room.

There is a question here of privacy. What do you think? Me? If I do not want people to see what I am doing in my apartment, I close the blinds. The light in her bedroom went out. She walked into the living room. She took Oscar by the hand and led him back into her bedroom. We filmed all of this. Invasion of privacy? We didn't care. We had caught Oscar red-handed, and it was likely this case would never go to court.

About 20 minutes later, Oscar emerged from the bedroom into the living room. The woman trailed behind, wearing a robe. He turned at the door, gave her a hug and kiss, and left. We filmed all of this, as we did when Oscar emerged from the building, got into his car, and drove off. Again, we did not follow him. It did not matter at that point. We had the second element needed in North Carolina to prove adultery: opportunity.

Was Oscar with a woman, not his wife, long enough for a sexual activity to have taken place? An unqualified yes. I called Nancy, told her Oscar was on his way home. I asked her to note the time he arrived. I asked her to discuss the game with Oscar to see how much he knew in case she had to testify.

We gave our film, videos, and reports to Nancy's attorney. What happened?

I cannot tell you. My partner and I were never called to testify. My job was completed.

I do know that Oscar sold his share of his business and moved out of North Carolina.

Personal Injury   These are civil cases, which mean one person is suing another person for a perceived wrong. Each party hires an attorney. The state, other than providing a judge, a courtroom, and a jury, if necessary, has almost no interest in the outcome. Most personal injury cases involve accidents such as automobile crashes. Who was at fault? Whose insurance company is liable?

The person who perceives himself or herself as the victim is suing for monetary compensation. Personal injury cases almost always involve a fight over responsibility, negligence, and the size of the award to the victim.

Personal injury cases might include workmen's compensation cases. A person falls at work, goes on disability because of a bad back, and claims he or she cannot go back to work. Most of these cases are legitimate, but there are those rascals who want to cheat. They want the free money even though there is nothing wrong with them. I might be hired to do a surveillance.

Does the man go bowling every Friday night? Does he water ski, do yard work, or play racquetball? I look for and document behavior that clearly indicates fraud.

Personal injury cases begin with a phone call from an attorney, representing a client who has been injured. We arrange to meet. At the end of the meeting, I return to my office and type up my notes. I prepare a list of folks to be interviewed as known at the time. What do I want from these folks? How do I reach them? Usually, I send the list to the attorney to make sure it is complete and that it meets his or her legal needs. Jodi McMaster (2006) suggests designing a table (see **Table 9.1**).

Attorneys sometimes request background checks of the potential clients. They need to know if there are any questionable motives. They want to know if there are any landmines. The client may have filed prior complaints against other defendants. If so, the client may be a professional plaintiff. The client's prior medical records need to be obtained to see if there are preexisting conditions. The client can still recover if the accident aggravated a preexisting condition, but the attorney needs to know.

The following is an example of a personal injury case I investigated.

An attorney, let me call him Joe, called me in for a consultation. Joe's client had been badly burned at a Charlotte nightclub. Joe needed facts to build upon his personal injury suit against the owners of the bar and perhaps employees who may have been negligent.

Joe had police and fire reports. He also had pictures of the burns on his client. Joe knew that this particular bar closed out the evening by pouring some alcohol in a trough behind the bar, lighting the vapors, and having the female employees dance on top of the bar. His client was walking by this fire when her hair caught on fire.

Owners of the bar contended the woman bent over to light a cigarette, and that is when her hair caught on fire. They claim she had contributed to her injuries.

I had an official list of folks who worked the bar that night. Interviewing them was on my to-do list, but what about other employees who might have been there?

Table 9.1 Initial Witness List

Witness Name	Role	Contact Information	Areas to Investigate
Dick Brown	Client was talking on her cell phone when the accident occurred	(704) 567-8900	Their conversation, client's mental and physical state

Adapted from: McMaster 2006 (p. 67). *Civil interviewing and investigating for paralegals.* Upper Saddle River, NJ.

EXERCISE 9.2

Preparing for a Personal Injury Case

Activity: The students plan basic steps for an investigation.

Purpose: To apply basic investigative interviewing techniques.

Scenario: An attorney called for a consultation. His client was badly burned at a Charlotte nightclub. He needed facts to build upon his personal injury suit against the owners of the bar and perhaps employees who may have been negligent.

The attorney had police and fire reports. He also had pictures of the burns on his client.

The attorney knew that this particular bar closed out the evening by pouring some alcohol in a trough behind the bar, lighting the alcohol, and having the female employees dance on top of the bar. His client was walking by this fire when her hair caught on fire.

Owners of the bar contend the woman bent over to light a cigarette, and that is when her hair caught on fire. They claim she had contributed to her injuries.

There was an official list of folks who worked the bar that night. Interviewing them is on the to-do list, but what about other employees who might have been there?

Put together a list of things to do and people to interview. Then students should check their list with the list below.

Hint: While it might seem an obvious thing to do, her attorney does not want you to talk to the client.

After a thorough examination of the information I had, here is what my to-do list looked like:

1. Interview all employees on the list supplied by management to find out what they say normally happens toward the end of each night, and what they saw happen the night in question.
2. Go to the club covertly and find out what other employees might have been working that night or hanging out but who were not on the list supplied by management. Interview them. While there, make diagrams. Take measurements of the width of the bar and trough.
3. Find out who was there from public safety agencies. Who investigated the incident—police, fire, emergency management technicians? Interview them. Look at their notes and pictures. Had this bar been cited for any violations, for instance, overcrowding, serving underage patrons, serving folks already drunk?
4. Interview witnesses on police reports. Find additional witnesses not listed and interview them.

5. Find the video for the commercial movie that had been made about this chain of bars, which was on the Internet. Watch it. Is there any information in that movie that might be helpful?
6. Google the bar chain. Is there any information on the Internet that might be helpful?
7. Get copies of any management manuals containing guidelines for crowd control when the bar lighting finale was about to begin.

There are too many interviews to go through each. I will highlight two of them and suggest a few questions.

I started with the fire department. Firefighters, as a general rule, are co-operative. They have no axe to grind, no dog in the fight. I found the fire investigator and went to see him. The purpose of the interview was to gather information about the club's past record of fires or fire safety violations, the incident involving the victim, and the fire department's experience with lighting vapors.

After the introductions, I began with an open-ended question: "What can you tell me about the fire incident at the nightclub?"

I listened and took notes during the lengthy narrative. I did not interrupt; I just let the interview flow freely and accepted whatever the fire investigator was willing to tell me. I took notes on the information. On a separate pad, I listed the probing questions I needed for follow-up. The following are examples of follow-up probes:

1. Had this type of accident happened at this club before? Had there been any police citations prior to this incident at this club? Most commonly, clubs get cited for violating occupancy limitations. Were there any occupancy violations?
   Had the fire department fined or cited the owners for any violations?
   Had the bar ever been shut down? Why? For how long?
2. Did they have any witnesses who were not listed on the police report? If so, I would request a copy of those witness statements. Specifically, did they know who ignited the vapors from the alcohol? Who directed that person to light the vapors?
3. Did the bar have manuals on how to control the fire and safeguard the patrons? If so, did the fire department have a copy? Did the fire department suggest any changes?
4. What employees were working that night? Were there any employees hanging out who were not on duty?
5. How often had the bar been visited by fire inspectors? What for? Did they find anything improper?

I asked for pictures, videos, and reports.

I now had a "blueprint" for further investigations. I had a list of the women working the bar that night. I made sure that I was supplied real names, instead of Ginger or Spice, with addresses and phone numbers. Still, almost none of the contact information was current. I had to track down these employees, who moved frequently and forgot to notify their employers of their new addresses. Suffice it to say, none of these women talked to me. The common answer was that management told them to refer all questions to the corporate lawyer.

I went to the bar one night, pretending to be a patron. When I arrived, there were a few guys milling around outside. I asked them if it was safe to go in, I had heard there had been a fire recently. One of the men volunteered he was there the night of the fire. In fact, he worked the club as a bouncer but was there that night just having a good time. His name was not on the list of employees supplied by management. I got his name and phone number. I went inside.

My client stated she had been in the ladies room, and when she came out and walked back to her table, she came close enough to the bar that her hair caught on fire. How far was it from the ladies room to the bar? At what point in the bar had the fire been lit? Was it possible for our client to bend into the flames?

One of the defenses presented by management was that my client stopped, bent over, and used the fire to light a cigarette. I surreptitiously measured the length and width of the bar. I stood at the end of the bar so I could measure the width of the trough. My client would have had to have been 10-feet tall to lean over and reach the flames.

Besides, we had witnesses who knew her. All of them said she was a non-smoker.

I called the off-duty bouncer and set up an interview. We agreed to meet at his apartment, which meant, I had little control of the environment. He and a roommate were watching TV when I arrived.

"Is there a place we can go and talk privately?"

The roommate got up, turned off the television, and disappeared down a hallway. We sat side by side on a couch. I pulled out my tape recorder, turned it on, and put it down on the coffee table, which I pulled closer to the couch.

EXERCISE 9.3

Planning the Interview

Activity: The students incorporate what they have learned about interview objectives and questions into a new setting. The students should complete the exercise before checking the author's questions below.

Purpose: To apply basic interview techniques.

You call the off-duty bouncer, who said that he had been at the bar (see Exercise 9.2) and set up an interview. You agree to meet at his apartment, which means you have little control of the environment. He and a roommate are watching TV when you arrive.

1. What do you do to attempt to have some control over the setting?
2. What would be your objectives for this interview?
3. What questions would be relevant to the objectives?

My objectives were to learn and document what he stated was the club's policy surrounding lighting vapors, what he saw the night of the incident, and who else witnessed the accident. Here are my questions:

1. What are the names of other folks who might have witnessed the incident?
2. Would he testify if necessary?
3. What did he see? Did my client contribute to her injuries?
4. Did he have a vision problem, hearing problem, drug habit, or a criminal record?
5. Were patrons pushed away from the bar prior to the lighting?
6. Had anyone else interviewed him?
7. Why was he there that night if he was off duty?
8. Did he see my client earlier in the evening prior to the incident? Did she appear sober? Was she smoking at her table?
9. Did he see her leave to go to the ladies' room?
10. Did he see her come out?
11. Could he tell with specificity what path she took when she left the ladies room?
12. Did she lean into the fire? If not, how close to the fire did she walk?

After the introductions, I looked around the apartment for a rapport builder. He had a college banner on the wall. Football is always a great way to start a conversation. Once the chitchat ended, it was time to get down to business. The first question was open ended, "Tell me what happened the night of the incident." Again, in this instance, I let him tell his narrative without interruption. I had the tape running, I was taking written notes, and on a separate paper, I kept track of probing, or follow-up questions.

I now had helpful information for my client and a more thorough blueprint for further investigations.

Most of you want to know how this case ended. I cannot tell you. As I have written before, I rarely follow my cases to conclusion. I do know this. The case was settled without going to court. I heard my client was happy with her monetary award.

Criminal Defense   All criminal cases involve a charge by the state or some branch of the government, against an individual. Criminal cases are easy to distinguish from civil cases. If there is a possibility the defendant might go to jail, it is a criminal case.

There is always someone representing the government on behalf of the people of the state. It might be a district attorney. It might be a federal prosecutor. In these cases, the defendant needs representation. An attorney is appointed by the court if he or she is indigent. If the defendant has money, he or she retains an attorney specializing in criminal matters.

These cases always begin with a phone call from an attorney representing a client. I arrange a meeting with the attorney. If the client is not in jail, I ask that the client be immediately available after that meeting. Generally, the attorney has already interviewed the client and has made extensive notes.

When I arrive, the attorney goes over his or her notes. We discuss areas to be investigated for corroboration or impeachment. Grounds for impeaching a witness are related to truthfulness. If the witness is for the state, we will be examining areas in which the witness can be shown to be less than truthful.

Then, the attorney gets the accused, brings him or her into the conference room, makes the introduction, and leaves me alone with the client.

Again, here is a real case. The facts and locations are real. The names have been changed for reasons of confidentiality. My client had been charged with various counts of sexual offenses. If convicted, he faced a number of years in prison.

EXERCISE 9.4

Preparing for a Criminal Defense Case

Activity: The students plan basic steps for an investigation.

Purpose: To apply basic investigative interviewing techniques.

Scenario: You have a client, Vic, charged with numerous counts of sexual offenses. There are no witnesses to the actual events in the car; however, there is a lot of information to be gathered through interviewing that could help your client.

Vic goes to the City Tavern, a restaurant in the Stonecrest Shopping Center. While he is there having a few drinks, Lisa, a short, blonde woman, is at the other end of the bar. They meet. They talk. They leave City Tavern and drive to Mickey and Mooch, a nearby restaurant with a more varied menu. Vic drives his car. Lisa leaves her car at City Tavern. They eat dinner at Mickey and Mooch. They have a few drinks, dance, and leave. Lisa is a little wobbly; nonetheless, Vic lets her drive his car. He's going to take her home because he does not want her to drive herself home in her car.

Up to this point in the narrative, there are no factual disputes between Vic and Lisa. Lisa says Vic told her to pull to the side of a road not far from her house. Vic says it was Lisa's idea to pull over. Lisa says Vic attacked her, pulled up her blouse, tried to have intercourse, and then ejaculated on her blouse. Vic says Lisa was the aggressor, and he ejaculated on her blouse because of her manipulations.

1. Figure out who the key witnesses are.
2. What do you want to know; what are the objectives of each interview?
3. What documents can you get?
4. What details of both locations do you need?

Vic was sitting in the conference room with the attorney when I walked in. Greetings and introductions followed. The attorney gave me a brief outline of the case, the charges, and the files he got from the district attorney's office, and then left me to interview Vic.

Understand this: Vic can be compared to a drowning victim. He views me as the person who is about to throw him a flotation device. He is eager and willing to talk. For the moment and the next few months, I am his only hope to keep him from prison. The attorney gets involved later after I do my job.

Here is the outline of the case:

It is a Thursday night. Vic is recently married. He is a landscape architect in a struggling business. His new wife has some wealth. On this night, Vic's wife is out playing cards with her female friends, girls' night out, a weekly ritual.

Vic goes to the City Tavern, a restaurant in the Stonecrest Shopping Center. While he is there having a few drinks, Lisa, a short, blonde woman, is at the other end of the bar. They meet. They talk. They leave City Tavern and drive to Mickey and Mooch, a nearby restaurant with a more varied menu. Vic drives his car. Lisa leaves her car at City Tavern.

They eat dinner at Mickey and Mooch. They have a few drinks, dance, and leave. Lisa is a little wobbly; nonetheless, Vic lets her drive his car. He is going to ride along as a passenger to make sure she gets home safely. I never found out what happened to Lisa's car at City Tavern.

Up to this point in the narrative, there are no factual disputes between Vic and Lisa. Here is where their narratives diverge. It is a classic nightmare for prosecutors and defense attorneys—he said, she said.

Lisa says Vic told her to pull to the side of a road not far from her house.

Vic says it was Lisa's idea to pull over.

Lisa says Vic attacked her, pulled up her blouse, tried to have intercourse, and then ejaculated on her blouse.

Vic says Lisa was the aggressor, and that he ejaculated on her blouse because of her manipulations.

When things calmed down, Vic says he wanted to take Lisa home, but Lisa refused to tell him where she lived. Lisa got out of the car and walked off. Vic let her go.

There are a couple of loose ends here that really do not play into the investigation.

First, inquisitive minds might want to know, why was Vic out trolling if he had just gotten married? I never asked Vic that question. Whatever, the answer was not relevant to the investigation. Second, what happened to Lisa's car? I never found out, and it was irrelevant to the investigation.

Two people served Vic and Lisa at City Tavern. Two people served Vic and Lisa at Mickey and Mooch. Lisa did go to the hospital, and she did report the crime to the police. Vic was arrested. His bride had to go to the jail and post his bond.

I started at City Tavern and interviewed the server who began waiting on Vic. I asked a number of specific, even leading or closed questions.

Is Vic a regular? Do you know him? Has he picked up women before? What time did he come in the night of the incident? Where was he sitting? What was he drinking? Was Lisa there when Vic got there? Do you know her? Is she a regular?

Vic got to the bar about 4 P.M., sat at the bar, and was drinking by himself. He was a regular; most of the staff knew him on sight. Lisa came in after Vic, about 6 P.M. Vic was sitting at the left of the bar; Lisa, on a stool seven seats to his right. Lisa was not a regular. Both servers thought it was Lisa's first time at City Tavern.

Lisa went to the restroom, and when she came back, she sat on the stool immediately to Vic's right. That is important. It shows, at a minimum, that Vic was not the aggressor.

City Tavern was getting busy. The bartender had other customers, so a second server took over. My questions for him:

Did you hear any conversation? What did they drink? Who paid? How did they appear to be getting along? Do you know Vic? Do you know Lisa?

At some point, close to 8 P.M., Vic and Lisa left. They walked out together and seemed to be enjoying each other's company.

Who paid for the drinks? The manager at City Tavern said he would find the credit card receipts and give them to me. I went back a week later and got them.

Vic paid for all the drinks. We now had a complete list of the drinks and how much Vic spent, and the two servers were able to remember with re-markably clarity what Vic drank and what Lisa drank.

When I came back, I took pictures of the bar, the stools, and the seating arrangements. I took measurements so my attorney would have a clear rep-resentation of the layout. With my notes, he could recreate the encounter.

Why did they leave City Tavern? Lisa, it turns out, had just moved to Charlotte. She had told Vic she was hungry, and she wanted to eat some-thing that Vic knew Mickey and Mooch specifically had on their menu. So they went there for dinner.

Next stop is obvious. Go to Mickey and Mooch. First, I interviewed the bartender who began serving them. I asked him the same questions I asked the bartender at City Tavern. These were pointed, directed questions. I needed specific answers to critical information. Open-ended questions, in this instance, would not be useful. I did a lot of probing during the inter-views, and all of them were tape recorded. I also got the receipts for food and drinks. Vic paid with the same credit card he used at City Tavern.

Mickey and Mooch has a place where people can dance. According to the bartender and server, Vic and Lisa danced, and she seemed to be "touchy feely" with Vic. At the bar, Lisa seemed to be playful and intimate, touching and rubbing Vic.

After dinner, they left. The bartender and server observed them walking out holding hands, shoulder to shoulder, although the server said he thought Lisa looked wobbly. None of this information proves anything, but in a cir-cumstantial spotlight, it sure could look to a jury that Lisa was open to a sexual encounter with Vic.

What did Lisa do when she got home? The police reports included inter-views with her parents. They stated that Lisa came home after midnight, went upstairs, stumbled and fell on the stairs, got up, and went into her room.

She did not tell her parents that she had been attacked. She took a shower. That afternoon she told her parents that she had been attacked. She went to the hospital for a rape examination. Lisa had some bruising on her right arm. She said Vic had grabbed her, which is possible. It is also possible that a seat belt could have caused the bruising. More likely, Lisa could have been bruised when she stumbled on the stairs. There was no evidence of rape. Lisa's blouse did have semen stains, although I do not think the police ever did a DNA analysis. After all, Vic never denied that he ejaculated on Lisa's blouse.

I met with Vic at City Tavern. We talked about his meeting with Lisa. We drove to Mickey and Mooch. I measured distances and stops at traffic lights, wanting a feel for that evening. We went into Mickey and Mooch. Again, I got Vic to talk about what happened there. I wanted to measure his answers against the information I got from the servers. I asked very pointed, challenging questions. When Vic forgot some details, I cynically challenged him on his selective memory.

We left Mickey and Mooch and drove the route to where they eventually pulled over. Again, I measured distances and stops at traffic lights. I wanted a feel for the evening. I wanted a timeline that could be overlaid against the timeline I got from the four servers.

I never interviewed Lisa. I never met her. As I said earlier, a case like this is difficult for both the prosecution and defense. Only two people know with certainty what happened in that car when Vic and Lisa pulled over. There is not an investigator in the world who can verify either version of events once they left Mickey and Mooch. That is what trials are for. Can the prosecution prove to 12 average citizens, beyond a reasonable doubt, that Vic attacked Lisa against her will?

Keep in mind the questions I ask, the way they are framed, and the responses. A good interviewer must view the client's and witnesses' versions through the eyes of an imaginary jury. I want information available to answer questions that a jury deciding Vic's fate would have.

Eventually, the ADA dismissed all the charges against Vic. At this writing, Vic and his wife are still married. His landscape business is struggling. And, they have a baby.

Conclusion

Defense attorneys need PIs just as prosecutors need police detectives. PIs actually are involved in more diverse types of cases than most police detectives. Attorneys need them to track down information that corroborates or contradicts what their clients tell them in criminal and civil cases.

Working with attorneys, PIs decide the importance of admissible evidence, especially in criminal cases. They systematically organize their investigations and interviews to provide additional information.

The types of cases that PIs investigate can be categorized broadly as domestic, personal injury, and criminal defense. Because PIs do not have the same legitimate authority that police detectives possess, they must be creative in rapport building and interviewing. Their interviews include people from all walks of life — from professionals, such as firefighters, police officers, teachers and doctors, to bartenders, neighbors, and street people.

PIs are required to be honest, accurate, thorough, and above all, ethical. Having said that, ethics can be argued until the cows come home, and we all cannot agree on what they are. There is that slippery slope some try to climb, while others just go around.

The reputation PIs earn early in their careers influences their relationship with attorneys, prosecutors, judges, and a variety of professionals with whom they regularly interact.

EXERCISE 9.5

Ethical Discussions

Activity: The instructor will facilitate discussion with the students surrounding the listed questions.

Purpose: To highlight the ethical issues surrounding PI work

1. What are the ethical issues surrounding individuals' privacy?
   a. Do students think that it is ethical to video interactions inside of houses that can be viewed from outside?
   b. Do students believe it is unethical for a spouse to put software in his or her partner's computer that allows the spouse to read the partner's email? Doing so is a federal crime, but the spouse believes his or her partner is cheating. What ethical dilemma does it place on a PI whose client is the spouse? In other words, the spouse comes to the PI and shows the PI evidence of his or her partner cheating based on email communication that was obtained illegally through reading his or her partner's email communication.
   c. Do students believe that ethics surrounding privacy issues differ based on different circumstances?
2. The PI misrepresents himself or herself as a patron rather than as a PI. The PI goes into the public place, buys a drink or a meal, and casually asks questions. Are there ethical issues surrounding a PI "snooping" around a public place such as a nightclub, bar, or restaurant without telling employees or the manager? Do you think the PI would have gotten information if he or she had identified himself or herself as a PI?

10
CHAPTER

Interrogation

OBJECTIVES   Upon completion of this chapter, the student should be able to

1. Distinguish between interviewing and interrogation
2. Discuss some of the major interrogation approaches
3. Understand the legal framework surrounding interrogations
4. Comprehend strategies used in interrogations

Introduction

The purpose of this book is to provide students the foundation to conduct different types of interviews that will result in obtaining valid, complete, and relevant information. Interrogations, no matter the definition, are used in the criminal justice setting to question potential suspects of crimes. Given the variety of types of interviews, interrogations represent only a small percentage of their use; however, the seriousness of interrogations' potential consequences requires some attention, and so, a separate chapter. The reader is encouraged also to review the ethical discussions in Chapter 2.

There are numerous volumes and a growing body of research written about interrogation. For those readers who aspire to become detectives in law enforcement agencies, it is advised they fully understand the implications and potential power of interrogation techniques.

Interrogation Defined

Devilish Rutledge (1996) defines interrogation as "controlled questioning calculated to discover and confirm the truth from the responses of an individual in spite of his intentions and efforts to conceal it" (p. 6). The purpose of most interrogations is to solicit incriminating evidence.

Interviews and interrogations commonly are viewed as different. Inbau, Reid, and Buckley (2005) distinguish between interviewing and interrogation. They describe interviews as conducted for the purpose of gathering information from individuals who have not been accused of some criminal activity. On the other hand, interrogations are designed to persuade suspects to tell the truth surrounding their involvement in criminal activity.

Custodial interrogations are legally differentiated from other interrogations and will be discussed in detail further in the chapter. In custody, the suspect is considered detained and must be read *Miranda* warnings.

In *Rhode Island v. Innis* (1980), the courts stated that interrogation is not only questions but also words or actions by the police that the police know are likely to extract an incriminating response. In *Rhode Island v. Innis*, the defendant had been arrested for murder of a taxicab driver with a sawed-off shotgun. Within the hearing range of the defendant, two officers were talking to each other. They said they hoped to find the gun that was used before some children found it. The defendant told the officers where the weapon was. The courts considered the statement admissible because the exchange was between officers and not directed toward the suspect nor were they provoking him to talk.

Precautionary Measures

Although it is a common myth that an innocent, mentally stable person cannot be influenced to confess to a crime, Chapter 2 describes experiments conducted by S. M. Kassin (1997, 2005) that support how easily even educated individuals can be manipulated to confess to an act they did not do and to actually internalize that confession. In dispute of the myth, more than 6,000 false confessions occur each year in the United States (Sear & Williamson, 1999). Also, while pleading guilty to a crime that a defendant did not commit is not the same as a confession, this unfortunate situation happens all too often. The end result is that innocent persons go to jail.

Confessions are one of the most powerful pieces of evidence that can be admitted into court. Gudjonsson (1993) cites several researchers who reveal that in only about half of the cases is the strength of collaborative or circumstantial evidence sufficiently strong against suspects prior to interrogation. It is a heavy ethical burden on investigators to thoroughly investigate a case before sitting a suspect down for an interrogation.

The British prefer to call interrogations investigative interviewing and emphasize the criterion is an open-minded search for the truth. Manipulative techniques used by U.S. officers are prohibited in Great Britain. The Police and Criminal Evidence Act (PACE) was implemented in 1984 to regulate interrogative practices. British detectives shifted from a prosecutorial orientation to one in which the search for the truth was uppermost (Sear & Williamson, 1999). They were to approach the investigation with an open mind, rather than a focus on interrogation of suspects that tested the suspects' statements against what detectives already knew or could establish. Strength of evidence in the case should encourage a switch of strategy away from focusing on confessions to obtaining more information from victims and witnesses.

The Justification Theme Development developed by the Reid School of Interviewing and Interrogation justifies or minimizes the moral seriousness of the criminal behavior allegedly conducted by the suspect (Sear & Williamson, 1999). The theme development often blames the victim, using such rationalization as "She was asking for it. What was she doing alone at a bar dressed like a vamp?" or "The company is rich. You were taking money to support your family." The philosophy behind theme development is that suspects' perceptions of their culpability will be reduced so they are more likely to confess. The long-term impact of minimizing the subjects' culpability is a serious consideration and will be further discussed in this chapter.

American courts also permit a certain degree of trickery and deception by the interrogator in attempt to gain a confession (Sear & Williamson, 1999). The belief is that any possible miscarriage of justice will be remedied through *due process.* At some point along the process, if there is mistaken identity of the suspect, the mistakes will be discovered. In most cases, confessions must be accompanied by corroborating evidence, even if circumstantial. In England, the defendant could be incarcerated on the confession alone, again pointing out the seriousness of the interrogation's consequences (Canter & Alison, 1999).

Inbau and colleagues begin their explanation of the Reid Nine Steps of Interrogation with three precautionary measures for the protection of the innocent (2004). Eyewitness identifications and motivations for false accusations continue to be a problem for the innocent. They suggest great caution when (1) victims initially state they could not identify the perpetrator; (2) there is a serious discrepancy between the victim's original description of the offender and the appearance of the identified suspect; (3) the victim identifies somebody else first; (4) other witnesses are unable to identify the offender; (5) there was limited opportunity to see the offender by anybody; or (6) considerable time lapses between the victim's view of offender and the identification of offender.

Even with these precautions, Joseph Abbitt was found guilty of two counts of first degree rape in 1995. Both victims individually identified Abbitt as their rapist when shown a photo lineup with color photographs of six subjects. Even though Abbitt's attorney introduced DNA evidence that excluded him, he was convicted and sentenced to two consecutive life sentences. In 2009, after 14 years in prison, Abbitt was freed. Additional DNA testing collected from one of the victims excluded Abbitt from the rapes (Hewlett and Young, 2009).

Inbau and colleagues (2004) also suggest caution when confronted with unsupported memories of past sexual or physical abuse that surface during adult psychotherapy as repressed memories. Another psychologist should be consulted.

The Psychology of Confessions

Gudjonsson (1993) reviewed a number of researchers' theoretical models of confession including the decision-making, cognitive-behavioral, and psychoanalytical models.

Hilgendorf and Irving (1981, as cited by Gudjonsson, 1993) developed the decision-making model of confession in which they argued that when suspects are interrogated, they are engaged in a complicated and demanding decision-making process. Some of their decisions include (1) whether to speak or remain silent, invoking their rights for an attorney; (2) whether to make self-incriminating statements; (3) whether to tell the truth or not; (4) if deciding to tell the truth, whether to tell the entire truth or only part of the truth; and (5) how to answer questions.

Hilgendorf and Irving note that decision making is governed by the suspect's willingness to gamble the odds of the possible consequences. The suspect balances the potential consequences against the perceived value of choosing a particular course of action. For some suspects, decision making is based also on factors related to self-esteem and social approval or disapproval. These factors can work in a number of ways. The suspect may deny guilt because admitting would sanction him or her with a socially disapproved act; however, not owning up to the crime might result in powerful self-loathing and social disapproval. The suspect might decide to repent, "getting it off of his or her chest," and accept the punishment, hopefully activating potential approval utilities. The police might attempt to manipulate the factors influencing the suspect's decision to talk to them and tell part or all of the truth. The suspect's feelings of competence and self-esteem are susceptible to manipulation. The police can manipulate the suspect's perceptions of the possible consequences by minimizing the seriousness of the alleged offense and magnifying the cost associated with denial and deception. Police also can impair the suspect's ability to logically process information and make a rational decision. The police utilize isolation, the suspect's existing level of anxiety and fear, and the suspect's unfamiliarity with and uncertainty of the situation; thereby forcing the suspect to confront private demons that may distort rational thinking.

Gudjonsson (1993) expands the decision-making model of confession into the cognitive-behavioral model of confession. Based on social learning theory, he includes antecedents, which are events that occur prior to the interrogation that facilitate a confession. Gudjonsson lists social (isolation); emotional (anxiety and distress); cognitive (assumptions, thoughts, innocence, coercion); situational (nature of arrest and familiarity with police procedures); and physiological antecedents. He then divides consequences into immediate and long term. Immediate consequences include police approval and praise, sense of relief and arousal reduction, need for attorney help, and confusion. Long-term consequences are public disapproval, shame, exposure, motivation toward reparation, judicial proceedings, and emotional arousal motivating the individual to return to a previous level.

So for example, the social antecedent of isolation might influence the suspect to accept the immediate consequence of police approval by confessing, but the confession leads to long-term public disapproval. Or, the emotional antecedent of anxiety and distress might lead to a confession so the suspect will have immediate emotional relief, but long-term consequences of guilt, shame, humiliation, and feelings of exposure on the one hand, and motivation toward reparation, or a form of penitence for doing wrong on the other.

Gudjonsson (1993) also relates the psychoanalytic model to confession. Simply described, people's behavior is controlled by the *id*, the childlike center that desires pleasure; the *superego*, the parental conscious; and the *ego*, the logical, objective computer-decision maker. Unconscious compulsion to confess is the superego's attempt at appeasement in order to settle the tension between the ego and the id. The suspect has an unconscious need to self-punish and has feelings of guilt. Only after the individual confesses does the ego begin to accept the emotional significance of the deed. People's knowledge of their transgressions produces a sense of guilt that is experienced as oppressive and depressing. Confession is cathartic and produces relief. At some point, long-term criminals, by virtue of early conditioning or habituation, no longer suffer pangs of conscious.

According to B. C. Jayne (1986), who collaborates with Inbau, Reid, and Buckley, a person chooses to confess when he or she believes the consequences of a confession are more desirable than the continued anxiety of deception. With the exception of criminals who have been so conditioned that they have lost all sense of their early social learning, the majority of people of all cultures experience some level of anxiety when they lie. They are directly taught that it is wrong to lie, and virtuous people tell the truth. They are told stories as children of bad people who lie and good people who are always truthful. Does everyone know the story of Pinocchio?

Just as children also learn that lying may keep them from getting caught and being punished, a deceptive subject's expectancy is that if he or she confesses, the consequences are inevitable, and the most desired goal is not to confess. So, it is the interrogator's goal to decrease the suspect's perception of the consequences of confessing while, at the same time, increasing the suspect's internal anxiety for lying. The interrogator attempts to change the suspect's expectancy through persuasion. To persuade effectively, the interrogator must be credible, sincere, and knowledgeable of the investigation and the suspect's background. The suspect's expectancy will change if he or she can be led to internalize a purpose, or a justification, to tell the truth, for example, "the parents would like to find their child's body so they can provide a proper funeral and burial service."

Deception　As described in Chapter 4, there are a number of cues that are reported to accompany deceptive behavior. Some nonverbal behaviors are under more voluntary control than others. Telling lies is likely to require more thought and is a more cognitively complex task than truth telling. The subjects' level of arousal and ability for cognitive processing influence their success in deceiving others. For example, subjects who are highly motivated to lie (perception of serious consequences) are more likely to fail in their attempts (Rudacille, 1998).

Edelmann (1999) studied a hypothesis that described lie tellers as insecure and concerned with impressions they are making, thereby appearing more guilty or anxious, more cognitively challenged, or more aroused than truth tellers. The rationale for the hypothesis was that most people tell lies less frequently than they tell the truth so they feel less confident and more insecure when they are trying to deceive. Consequently, people who lie attempt to exercise greater control over their behavior by planning their performance. Their deceptive behavior appears to others as planned and lacking in spontaneity. Exceptions are experienced and confident liars, especially those with the opportunity to plan their deceit.

Some nonverbal cues are more likely to show deception than others. Individuals can control their facial expressions more than the tension in their shoulders. They can control their hand activity more than their feet motion. Edelmann reports that lie tellers usually take longer to respond before speaking and then speak more slowly. He also lists such nonverbal cues as pupil dilation, rapid eye blinks, higher voice pitch, and speech errors.

Edelmann suggests that close examination of smiling actually reveals deception; however, these differences probably cannot be discerned by observers. When enjoyment actually is experienced, there is muscle activity around the orbits of the eyes, not just the lips turning up. When concealing strong negative emotions, the smile contains traces of the negative emotions of disgust, anger, or fear, so the lips briefly may turn down before up, or the chin may quiver.

People trying to be deceptive increase their eye contact, but these are subtle cues. Because of the physiological effect of arousal, subjects telling lies blink more, and their pupils dilate. Due to the emotions associated with lying, their voices are slightly higher pitched, hesitant, and overly rehearsed.

Lie telling is cognitively complex, and there is a need to focus attention on verbal content and on attempts to control the speaker's own actions. Illustrative hand movements, those movements that accompany accent and speech, decrease. Deceptive people tend to keep their hands still and close to their bodies, although some deceptive interviewees tend to exhibit more adapter movements such as grooming, rubbing, and scratching. The adapter behaviors provide breathing room while they form a response.

The more experienced the liar, the more likely the deception. A highly experienced, self-confident, and socially skilled liar who has taken the time to plan his or her lies, but is not highly motivated to deceive, will be erroneously assumed to be telling the truth.

A recent example of a complex deception is the Madoff fraud. Bernard Madoff, former chairman of the NASDAQ stock exchange and founder of the Wall Street firm Bernard L. Madoff Investment Securities LLC, instigated what may be the largest investor fraud case ever committed by a single person. An unnamed investor remarked, "The returns were just amazing and we trusted this guy for decades—if you wanted to take money out, you always got your check in a few days. That's why we were all so stunned" (Henriques, 2008).

Because lying is difficult, most alibis are shallow with few details. If the interviewer has to pull the information out of the interviewee, then deception of some sort could be occurring, even with victims.

"My truck was stolen!"

"What happened?"

"I don't know. I went in to visit my friend for awhile and when I came out, it was gone!"

As opposed to:

"My truck was stolen!"

"What happened?"

"I think somebody must have hot-wired it. I have a 2006 Ford Rancher, and I know I locked it because I left my laptop computer behind the seat. I parked on the curb right outside my friend, Josie's, apartment at about 6 P.M. Her apartment is on the backside so I didn't hear or see anything, but when I came out around 11 P.M. to go home, it was gone. I ran across the street to that pool hall and asked the owner if he had seen anybody near my truck—I described it, a black 2006 Ford Rancher—but he just shrugged and said he had been too busy. He let me use the phone to phone the police."

If the interviewee is just describing what occurred, the facts should be consistent; however, a deceptive person will create information as asked so his or her statement will often have information that seems inconsistent. As my mother always said, "If you are going to lie, you better have a good memory." A deceptive person must remember what he or she said, not what actually happened.

To stall for time to create further fabrication, the deceptive person often answers a question with a question:

Girlfriend: "You were supposed to pick me up an hour ago. What happened?!"

Boyfriend: "Are you sure we said 7 P.M.?"

Girlfriend: "Yes, you said that we would get a bite and then go to the 8:50 movie."

Boyfriend: "Well, we still have time for the movie."

Girlfriend: "You never told me where you've been!"

Boyfriend: "Am I supposed to tell you every time I breathe? Let's go before it gets any later!"

These conclusions must be interpreted cautiously. If subjects are placed in tense situations such as interrogations, the serious consequences of their response may make them anxious and slower to form their response.

Accuracy in experts' ability to detect deception ranges from 40% to 60% which is a percentage range comparable to chance (Meissner & Kassin, 2004). As discussed in Chapter 4, the interviewer must exhibit caution when attempting to interpret deception. The probability for detection is more likely to increase if the interviewer looks for change in baseline behavior. As a reminder, baseline behavior, or the behavior shown when the interviewee is describing nonthreatening subjects, is compared with changes in the interviewee's behavior when describing more threatening subjects.

Legal Framework of Interrogations

The interrogator should be guided by the Fourth, Fifth, Sixth and Fourteenth Amendments of the U. S. Constitution. The U.S. Supreme Court, through its decisions, condemns improper use of questioning. Based on these four amendments, before a confession is admissible (1) it must be given freely and voluntarily; (2) if in custody, the defendant must have been given *Miranda* warnings; (3) it must not be contaminated by an illegal search or arrest; (4) provision for counsel or the waiving of such provision should be provided, and (5) delay in arraignment requirements should be met (Klotter, Walker, & Hemmens, 2005).

Fourth Amendment

The Fourth Amendment applies to the need for suspects' admissions and confessions to be based on legal searches and seizures, or arrests. For example, if a confession is obtained as a direct result of an illegal arrest without probable cause, the confession is inadmissible in court because it was obtained in violation of the Fourth Amendment. The confession is considered "fruit of the poisonous tree" (*Wong Sun*, 1963).

In *Dunway v. New York* (1979), even though Dunway went with the police without any resistance, he was arrested without probable cause, and evidence revealed that he would have been restrained if he had attempted to leave. Dunway was picked up on charges of attempted robbery and murder based on information from an inmate. He confessed to the crime, but because the circumstances under which he was picked up were unlawful, the confession was considered illegally obtained.

Even if *Miranda* warnings are given, a confession that immediately follows an illegal arrest or search will be inadmissible if there is a causal connection between the illegal search or arrest and the confession (*Lanier v. South Carolina*, 1985). The federal and state courts continue to hear similar cases and apply the same standards for police conduct (Klotter et al., 2005).

Fifth Amendment

Based on this amendment, no person shall be compelled in any criminal case to be a witness against himself or herself. The two acceptable conditions are the presence of the defendant's attorney, or a defendant who, after being given the *Miranda* warnings, agrees to answer questions.

The *Miranda* warnings state that the suspect has a right to remain silent, any statement the suspect makes may be used as evidence against him or her, and the suspect has a right to the presence of an attorney either retained or appointed. The exercise of these rights must be honored, but the defendant can waive these Constitutional protections.

If the suspect chooses to waive his or her rights, the burden of proving that the waiver was knowingly, intelligently, and willingly given falls to the prosecution. In 1972, the Supreme Court determined that the Constitution required the lower standard "preponderance of the evidence" rather than "beyond a reasonable doubt" as the standard of proof (*Lego v. Twomey*, 1972, as cited in Klotter, Walker & Hemmens, 2005). Preponderance of the evidence is often measured by the weight of proof being slightly heavier on one side than the other. In the case of proving the defendant knowingly, intelligently, and willingly waived his or her rights, the prosecutor's evidence would need to weigh slightly heavier than the evidence for the defendant.

A voluntary statement can be inferred from the actions and words of the person interrogated; however, silence from the defendant after the warnings is not sufficient proof that he or she understood the warnings. To prevent any suggestion of any kind of impairment, it is useful to ask the suspect to tell, in his or her own words, what the rights mean and then record his or her comments. Likewise, there must be no implied threats or trickery/deception that influences the suspect to waive rights. If the suspect's rational choice is distorted in any way, the confession is considered involuntary (*Pyles v. State*, 1997, as cited in Klotter, Walker, & Hemmens, 2005).

Custody is an important criterion for requiring the *Miranda* warnings. Obviously, suspects are in custody if they are in a lock-up facility, locked in a police car, handcuffed, or told they are under arrest. The test is whether a reasonable person in the suspect's position would have felt that he or she was under arrest or being restricted to the degree associated with arrest.

Oregon v. Mathiason (1977), as well as several more recent state cases, note that simply holding the questioning in a police station is not sufficient to be considered in custody as long as the interviewee is told he or she is not under arrest, the room is unlocked, and there is no suggestion of detention. In the case of Mathiason, an officer who Mathiason knew called him and asked Mathiason to meet him at the police department. Mathiason agreed. The two shook hands when Mathiason arrived. Mathiason was told he was not under arrest, but the officer wanted to talk to him about a burglary that police believed he had committed. Mathiason confessed after a short period of time during which *Miranda* warnings were not given. Although the Oregon courts said the questioning was custodial, the U.S. Supreme Court reversed the decision and said it was not custodial; consent negates custody.

Often an individual is asked to come into the police department for an interview. At some point during the interview, information may be developed that turns the interview into a custodial interrogation. At that point in time, *Miranda* warnings should be given before questioning proceeds. The interviewer must be able to determine when that change occurs; the courts will certainly make such a determination.

The action of questioning is critical in determining the need for *Miranda* warnings. If a subject is in custody, but no questions are asked, *Miranda* warnings are not required. If the subject chooses to talk and even confess voluntarily, his or her statement is admissible in court.

Although officers asking a general question such as "What happened?" when arriving at a scene is acceptable, what is or is not questioning is not straightforward. The Christian Burial Speech case is a classic case that illustrates this point. In *Brewer v. Williams* (1974), after a child disappeared, Robert Williams was arrested. He was seen leaving the YMCA with a large bundle with what appeared to be two skinny legs protruding from the bag. Clothes that belonged to the missing child were found in his car.

While transporting Williams, the officers began talking to Williams about the need to give the child a Christian burial before a snowstorm prevented the child's body from being found. Williams agreed to take the officers to the body. The Supreme Court stated the officers' actions amounted to an interrogation in violation of the Fifth Amendment (Klotter et al., 2005). The decision of *Brewer v. Williams* contrasts sharply with *Rhode Island v. Innis*, 1980). As noted earlier, in *Rhode Island v. Innis* the officers were talking to each other and not directly to the defendant, so an interrogation did not take place. The Supreme Court noted that questioning under *Miranda* refers not only to express questioning, but actions by the police that they would know are reasonably likely to elicit incriminating responses from the suspect.

Sixth Amendment   This amendment of right to counsel is for postindictment interrogation, while the Fifth Amendment is protection pre-indictment. According to the Sixth Amendment, once a suspect has been formally charged, he or she can be questioned further only if the prosecution can prove that the suspect volunteered to talk and knowingly relinquished right to counsel.

Escobedo v. Illinois (1964) is considered the landmark case in which the right to counsel was extended to suspects during interrogation or other pretrial procedures. Danny Escobedo was arrested and taken to the police department for questioning in connection with the fatal shooting of his brother-in-law that had occurred about 11 days previously. Several days earlier, Escobedo had been arrested for the crime but had made no statement and was released after his lawyer obtained a writ of habeas corpus from a state court. During the second arrest period, Escobedo made several requests to see his lawyer. His lawyer was present in the building and made a persistent effort to see Escobedo but was refused access to his client. Escobedo was not advised of his right to remain silent and, after persistent questioning by the police, made a damaging statement, which was admitted at the trial. Convicted of murder, he appealed to the State Supreme Court, which affirmed the conviction. The U.S. Supreme Court reversed the conviction and declared it inadmissible, and designated the point during the process at which right to counsel must be allowed. When an investigation is no longer a general inquiry into an unsolved crime, but has begun to focus on a particular suspect in custody, he or she should be provided an opportunity to consult with an attorney. Based on the Sixth and Fourteenth Amendments, every defendant should, at that transitional point, be warned of his or her constitutional right to keep silent and be given the opportunity to have an attorney present. If these actions are not taken, then no statement extracted by the police during the interrogation may be used against him or her at trial.

Once right to counsel is invoked, the police cannot initiate any interrogative questions.

In *Edwards v. Arizona* (1981) the subject was advised of his constitutional rights to which he replied that he wanted an attorney before talking. Questioning stopped at that point in time. A little while later, two detectives came to talk to Edwards. Edwards was given the *Miranda* warnings again, following which he agreed to talk after he heard the taped statements of one of the accomplices. He then confessed and was convicted. The Supreme Court overturned the lower courts' decisions that allowed the confession to stand, stating that the police should not have talked to Edwards once he had expressed his desire for counsel unless Edwards initiated further communication with the police (Klotter et al., 2005). What has come to be called the Edwards rule applies even when the questioning by different officers, who were not aware that the suspect had requested an attorney, concerns an unrelated crime (*Arizona v. Roberson*, 1988).

Fourteenth Amendment

This amendment discusses due process, the voluntariness and lack of coercion of the questioning of the suspect. There cannot be any attempt to pressure the suspect or to overcome the suspect's will to resist being questioned.

To be admissible, there must be documentation that the confession was made voluntarily, without inducement or the slightest hope of benefiting from the confession, and with no remote possibility of threat of injury. If the suspect is impaired in such a way that the suspect cannot think for himself or herself, an interrogation should not take place. The suspect cannot be interrogated excessively or promised anything.

Any condition before or during interrogation of a suspect that might raise an issue of a voluntary confession should be addressed and documented. The interrogator should question the suspect, friends, or family about any possible impairment. Any incapacitating condition, such as the need for medical care, lack of sleep, or request for food or water, should be addressed.

Before beginning to ask any crime-related questions, the interrogator should ask questions that build rapport and demonstrate the suspect's ability to have a coherent, rational conversation. Questions with recorded answers about time orientation, when the suspect last slept, any medication or medical care currently received, education, and employment are relevant to determining the suspect's rationality.

Permissible Interrogation Tactics and Techniques

Trickery or deceit can be used in some circumstances. For example, in *Frazier v. Cupp* (1969), the confession was obtained after the defendant was told falsely that his suspected accomplice had confessed. The Supreme Court stated that the misrepresentation was insufficient to make the voluntary confession inadmissible.

A number of cases dealing with deception or trickery by the police have been reviewed by the state courts, who in general have concluded that trickery or deceit must not be of such a nature as to "shock the conscience" of the court or community, nor consist of an act that is apt to induce a false confession.

The police have told suspects that they want to talk to them about a lesser crime (*Colorado v. Spring*, 1987) or that they have physical evidence that incriminates the defendant (*State v. Kelekolio*, 1993). The courts upheld the confessions, finding that the false statements were not likely to produce untrustworthy confession.

Andrew Dalzell was suspected in the homicide of Susan Kilmer in North Carolina in 1997. Andrew was arrested in Lincoln County, North Carolina, for obtaining property by false pretense, financial identity fraud and possession of stolen property in Carrboro, a town three hours away from Lincoln County. He was not given his *Miranda* rights.

Officers placed Andrew Dalzell in the back of an unmarked police car for the trip back to Carrboro and put a fake arrest warrant for homicide on the seat beside him. One of the officers, while traveling back to the police department read a fake letter that stated authorities would seek the death penalty unless Dalzell told them where he disposed of Susan Kilmer's body.

Dalzell became upset during a rest-stop bathroom break. One of the officers encouraged Dalzell by saying, "Tell the truth about whatever happened. Be a man and let the demon go." Dalzell blurted out, "I didn't mean for it to happen. I just snapped and took her body to Wilmington" (Ataiyero, 2004)

Orange County Superior Court Judge Wade Barber suppressed the confession, stating the police violated state law by misleading Dalzell about the nature of the charges he faced and by delaying in giving him his *Miranda* rights.

James Jackson was requested on several occasions by police detectives to answer questions about the death of Leslie Hall-Kennedy. Although Jackson always was given his *Miranda* warnings, which he waived, he was never placed under arrest.

Jackson was always free to leave and go home. The detectives normally gave him a ride back home and let him out. During the investigation, a kitchen knife identical to the knife found missing from a set that the victim had owned was found in the area of the crime scene. One of the detectives obtained a similar knife, pricked his finger, and made a thumb print from the blood on the knife blade.

Police asked Jackson to come in for another interview. Again he agreed to be interviewed; he was not arrested; and he was free to go. Jackson was asked how he could explain his fingerprint on the knife. Jackson stated that it could not be his print.

The detective also said that a witness had come forward who saw Jackson running out of the victim's apartment (false evidence). The detective said that Jackson had friends in the interviewing room. To further deceive Jackson, the detective falsely said that a witness had come forward who saw Jackson running out of the victim's apartment. The detective then stated to Jackson that the detectives were Jackson's friends. They just saw inconsistencies between Jackson's earlier statements and the evidence they now had. One detective stated that they thought Jackson had committed the homicide, and what the detective wanted was a clarification of the discrepancies in Jackson's statements and the evidence.

The detective went through the evidence again including the fingerprint on the knife. Jackson admitted to touching the knife, but continued to deny killing the victim. The detective also deceptively described other physical evidence that actually had not been found. Jackson then told the detective that he had killed the victim. The detective said that he knew it and wanted to know why Jackson had killed her. Jackson then told the detective his version of what had occurred and that he stabbed the victim (*State v. Jackson*, 1983).

Did the courts suppress Jackson's confession? Yes and No

After the trial court denied the confession, the state appealed the trial court's decision to suppress. The N.C. Supreme Court reversed the trial court's decision. The Supreme Court held that (1) evidence supports that the defendant was not in custody prior to his confession; (2) evidence supports the finding that the defendant's confession was voluntary even though the detectives deceived and lied to him; and (3) the defendant's confession was admissible under the N.C. test to determine admissibility. Jackson's confession was voluntary under the totality of circumstances.

Emergency Exception to *Miranda*

There is a life-saving–rescue emergency exception to the *Miranda* warning mandate.

Because the public interest outweighs private interest, in cases in which the possibility of loss of life can be prevented or time is critical, a *Miranda* warning is not required. Rescue must be the primary purpose and motive of the interrogation.

> The police capture the man who they believe kidnapped a teenager. The ransom note to her parents stated, "You have eight hours to get me the $1 million. I have buried your daughter in a wooden coffin. There is sufficient air for her to live eight hours." The police don't waste time with the *Miranda* warning. They ask the man, "Where did you bury the girl?"

(For more of this fictionalized version of the absence of a *Miranda* warning, watch the original *Dirty Harry* movie.)

Following is another example:

> A man was arrested and charged with rape.
>
> The victim stated that the man had put a gun to her head and threatened to kill her. The man was found hiding inside a grocery store. He was handcuffed outside the store and searched. He was wearing an empty shoulder holster. Without being given the *Miranda* warnings, he was asked what he had done with the gun. The man took them to the gun hidden inside the store.
>
> The courts considered the subject's actions admissible in court, and the public was safe from the weapon (*New York v. Quarles*, 1984).

Youthful Suspects

As with all interrogations, the issues of voluntariness and the totality of circumstances are critical. The age of the suspect adds another dimension to the ability to knowingly and willingly waive rights. Some courts do not allow waivers by youth under a certain age unless the youth has received consultation with an interested, informed, and independent adult, usually a parent or an attorney retained by the family.

An example of a police agency's standard operating procedures as they relate to juveniles specifically states that juveniles under the age of 14 years old must have their parent, guardian, custodian, or attorney present. It also states that no more than two officers may simultaneously participate in interrogating a juvenile. There is also a clause that discusses the length of the interrogation be based on the "seriousness of the crime, the juvenile's age and maturity level and the juvenile's previous experience with law enforcement" (Charlotte Mecklenburg Directives, Police Department, 2005).

Beyond a specified chronological age requirement, the courts also consider education, background, and intelligence of a defendant as an emotional or mental age of functioning. The degree and extent to which the suspect possesses the capacity to understand his or her predicament and to relate with reliability the happening of an event will depend upon the totality of the circumstances.

For example, on December 15, 1995, Jose Rosario Garibay, Jr. was arrested crossing the Mexico–United States border after U.S. custom agents discovered more than 100 pounds of marijuana in the trunk of his car (*United States v. Garibay*, 1998). The U.S. Court of Appeals overturned the lower court's decision to deny his motion to suppress his confession, and reversed the conviction of importing marijuana into the United States and possession of marijuana with intent to distribute. The U.S. Court of Appeals found that under the totality of the circumstances, Jose Garibay had not made a knowing and intelligent waiver of his *Miranda* rights. Although his primary language was Spanish, Mr. Garibay's rights were given to him in English, he was borderline retarded, and he did not sign a written waiver. Without his confession, there was insufficient evidence to support the conviction.

In another case, Marvin and Archie Cooper were convicted of armed robbery in state court (*Cooper and Cooper v. Chatham County*, 1972). Although several witnesses testified that the Cooper brothers had been retarded since birth, the district court in Chatham County, Georgia, declared they understood the *Miranda* warnings administered to them, and therefore, their confessions were voluntary and not in violation of their constitutional rights. On appeal, the court reversed and held that the brothers were mentally retarded and were not capable of comprehending the *Miranda* warnings. The court took into consideration petitioners' mental deficiency, age, and lack of familiarity with the criminal process. It concluded that the brothers' *Miranda* waiver was not knowingly and intelligently provided.

Major Interrogation Strategies

Important Principles   The interrogation of suspects should be one of the last activities of an investigation with all of the evidence and interviews of victims and witnesses pointing at one specific suspect(s).

The detectives should be clear about the method used to commit the crime, what the perpetrator would need to have done or known in order to commit the crime, the elimination of the possibility of a false or fraudulent report, and any possibility of an inside job. The suspect's alibi should be thoroughly checked prior to any interrogation. Using interrogations to narrow the field of potential suspects is questionable investigative behavior.

Dissonance is the stressful state in which the mind is not in equilibrium (Zulawski & Wicklander, 1993). Initially for all individuals, there is internal stress between the knowledge of what is right and deception, which violates what their parents taught them and society's moral code. As an individual lies and avoids getting caught, he or she experiences less anxiety and builds more rationalization for his or her crimes.

As noted earlier, the decision to lie is based upon the individual's assessment of whether the benefits of telling the truth outweigh the consequences resulting from an admission or confession. The decision to confess is to a large extent determined by the benefits perceived by the individual. For suspects, the consequences can be personal embarrassment, loss of a job, prosecution, and loss of freedom. Sometimes the guilt they are feeling is sufficient, or they cannot stand up against perceived overwhelming evidence. Suspects have to try to decide what is in their best interest: invoke their rights or waive them. If they waive their rights, do they tell the truth or lie? If they lie, they then attempt to prevent a successful interrogation or misdirect to create a miscalculation.

Interrogations are connected to the potential criminal prosecution of the accused. The interrogator must be knowledgeable of the elements of the crime that is being investigated. The goal is to identify the correct offender(s) and to obtain a confession(s); that is, a statement that details the involvement in the crime and all of the elements of the crime as well as the intent, motive, and offenders' mental state of mind.

The interrogator does not want to settle for an admission if possible. An admission is a statement in which other evidence is required to prove guilt. For example, the suspect admits to being at the crime scene when the crime happened but denies involvement in the actual crime. The suspect has admitted to the opportunity but not to the actual commission.

The interrogator should anticipate likely defenses and disclaimers from the suspect. The suspect might claim that the homicide was in self-defense, a mistake or accident, or in the heat of passion. The suspect might claim that he had sexual intercourse with the victim, but the act was consensual, not rape (Rutledge, 1996).

Introductory Statements

Inbau and colleagues (2005) recommend that the interrogator gets the suspect conditioned to telling the truth by starting with neutral areas and a relaxed manner: "We have plenty of time to talk. I want to hear what you have to say, but first I would like to spend a few minutes getting to know you . . ."

Miranda rights must be read if the suspect is in custody, so when possible, the interviewer should attempt to make the interrogation consensual and notify the suspect that he or she is not in custody.

The request for the interrogation should be casual if possible: "We would just like to clarify some information," or "We are trying to eliminate a number of people by talking to them." The detectives should not lie, but they do not need to tell the suspect that he or she is the prime suspect. Police now use the phrase "a person of interest."

In setting the stage for truth telling, the detectives should emphasize that their roles are to establish the truth. They want to convince the suspect that they have thoroughly investigated the crime: "Through extensive investigation, we already know a great deal. The important thing is that you tell the complete truth. We want to hear your side."

Interrogation Approaches

There are a variety of different interrogative approaches, all of which have their strengths and weaknesses and work under certain circumstances. Rabon (1992) and Zulawski and Wicklander (1993) describe the factual and indirect approaches. Probably one of the approaches best known is the Reid Nine Steps of Interrogation (Inbau et al., 2004; 2005).

Factual Approach   The factual, or direct, approach is used when the suspect has provided an alibi, and the investigator demonstrates through evidence collected and analyzed that the alibi can be impeached. It also is more successful when the subjects are first-time offenders, and their alleged crimes are precipitated by emotions such as passion, anger, or jealousy.

Using open-ended questions, the suspect is encouraged to tell what happened. The interviewer remains interested and encouraging. As the suspect grows more confident, his or her story will become more elaborate. After the suspect has been encouraged to expand as much as possible on the alibi, the interviewer presents evidence that puts holes in the alibi's credibility and convinces the suspect he or she is directly tied to the crime.

The testimony of Captain Jeffrey MacDonald provides an excellent example of the factual approach. Jeffrey MacDonald developed an elaborate story about the death of his wife and two young children. The police were able to find large, gaping holes in his alibi with the physical evidence.

As discussed in detail in Chapter 6, the early morning hours of February 17, 1970, the dispatchers on the military base of Fort Bragg, North Carolina, received an emergency telephone call from Jeffrey MacDonald, a military physician, "My wife and two daughters have been murdered." When the military police arrive at the MacDonald's residence, MacDonald was found alive but wounded next to his murdered wife in their bedroom. His most serious wound was a small, sharp incision that caused a lung to partially collapse. He was released from the hospital the next week.

MacDonald was able to provide the police a detailed account of the attack before he was taken to the hospital. MacDonald said that he had fallen asleep on the couch and was awakened by the screams from his wife and older daughter, Kimberly. He was adamant that there were four people in his house who killed his family and attacked him, leaving him unconscious.

MacDonald's wife, Colette, was stabbed 20 times in the chest, her skull was fractured, and both of her arms were broken, probably from attempts to fend off the attacker. Two-year-old Kristen and five-year-old Kimberly were beaten and stabbed multiple times.

Please refer to **Table 10.1**. The left column contains portions of Captain MacDonald's statement, and the right column describes the evidence that was discovered and analyzed.

Table 10.1 Jeffrey McDonald's Statement Compared to the Evidence

Statement	Evidence
I was hit the first time, and the guy hit me in the head. So I was knocked back on the couch, and then I started **struggling** to get up, and I could hear it all then—now I could—maybe it's really, you know—I don't know if I was repeating to myself what she just said or if I kept hearing it, but I kept—I heard, you know, "Acid is groovy. Kill the pigs." And I started to **struggle up**; and I noticed three men now; and I think the girl was kind of behind them, either on the stairs or at the foot of the couch behind them. And the guy on my left was a colored man, and he hit me again; but at the same time, know, I was kind of **struggling**. And these two men, I thought, were punching me at the time. Then I—I remember thinking to myself that—see, I work out with the boxing gloves sometimes. I was then—and I kept— "Geez, that guy throws a hell of a punch," because he punched me in the chest, and I got this terrific pain in my chest. And so, I was **struggling**, and I got hit on the shoulder or the side of the head again.	The disorder was minimal in the living room, where MacDonald describes all of his struggling with three men. Among the magazines spilled on the floor was an edition of *Esquire* that featured the Charles Manson-led murders. The police detectives quickly saw similarities between MacDonald's account and the stories of violent hippies high on hallucinogenic drugs. No blue threads that matched MacDonald's pajamas were found in the living room where he claimed to have fought; however, 80 fibers were found in the master bedroom and 19 in Kimberly's room including one fiber under Kimberly's fingernail.
And so, I let go of the club; and I was grappling with him and I was holding his hand in my hand. And I saw, you know, a blade. I didn't know what it was; I just saw something that looked like a blade at the time. And so, then I concentrated on him. We were kind of struggling in the hallway right there at the end of the couch.	MacDonald's blood is only found by the kitchen cabinet that contained rubber surgical glove's and in the bathroom sink. The police believe that the sink is where he made a surgical cut to his own chest. None of MacDonald's blood was found in the hallway where he stated he had been stabbed. The only trace of his blood in the living room was on his glasses and the copy of *Esquire* magazine.
I got a glimpse was, was some stripes. I told you, I think, they were E6 stripes. There was one bottom rocker and it was an army jacket, and that man was a colored man, and the two men, other men, were white.	Jeffrey MacDonald had very poor eyesight yet he was able to see detail that evening without his glasses. The house was pitch black when the police arrived, yet he described with some detail the three men and one woman.
The kitchen light was on, and I saw some people at the foot of the bed and I went down and –to the bedroom. And I had this—I was dizzy, you know. I wasn't really—real alert; and I—my wife was lying on the—the floor next to the bed.	As noted earlier, when the police arrived, the house was pitch black. MacDonald never mentioned turning a light off, and, strangely enough, he never turned a light on, even to phone the police or check on his family.

(Continued)

Table 10.1 Jeffrey McDonald's Statement Compared to the Evidence (*Continued*)

Statement	Evidence
And there were—there was a knife in her upper chest. So, I took that out; and I tried to give her artificial respiration but the air was coming out of her chest. So, I went and checked the kids; and—just a minute—and they were—had a lot of—there was a lot of blood around. So, I went back into the bedroom.	Each family member had a different ABO blood group so it was possible to track victims based on blood group.
	Among the blood-stained sheets near where Collette was found was the finger section of a latex surgical glove. It was torn as if taken off in a hurry.
	None of his blood or fingerprints were found on the knife that he claimed to have pulled from Collette. A knife and ice pick were found in the backyard, wiped clean.
	Blue fibers that matched MacDonald's pajamas were found under Collette. Eighty more fibers that matched his pajamas were recovered in the master bedroom. Nineteen blue fibers were found in Kimberly's bedroom including one under her fingernail.

Note: Jeffrey McDonald's statement to the detectives is part of his trial transcript as is the physical evidence found (*United States of America v. Jeffrey R. McDonald*, 1979).

Liars must rely on their imagination because they do not have any actual memory of the event as they say it happened. The main facts of their alibi will remain consistent, but they do not pay attention to small peripheral details. After relaying their story, interrogators should ask detailed questions that might appear irrelevant. For example, "You mentioned that you were inside the club with your friend Pete when the victim was shot. What kind of music did the club have that night?" If the suspect is responsible for a drive-by shooting of a victim in the parking lot, it is those kinds of questions that will help poke holes in the alibi and shake the suspect's confidence.

After giving his statement, the detectives asked Captain Jeffrey MacDonald some follow-up questions. MacDonald attempted to distract by denying specific details around the murders of his wife and children:

> McDonald: "Well, what, what are you trying to say?"
> Investigator: "That this is a staged scene."
> McDonald: "You mean that I staged the scene?" (repeats, but does not deny)
> Investigator: "That's what I think."
> McDonald: "Do you think that I would stand the pot up if I staged the scene?" (question, but still no denial)
> Investigator: "Somebody stood it up like that."
> McDonald: "Well, I don't see the reasoning behind that. You just told me I was college-educated and very intelligent." (distraction, but no denial) (McGinnis, 1983, p. 131).

In cases with multiple perpetrators with varying involvements, the interrogator initially can induce individuals with peripheral involvement to cooperate and provide information about the key players.

There are some risks to the factual approach. It precludes any possibility of rapport building or obtaining additional information. It must be a strong case, or the interrogator runs the risk of the suspect recognizing the case's weaknesses and not confessing. There is the risk also of the suspect immediately invoking his or her rights or flatly denying involvement, especially if the alibi has not been fully developed.

Indirect Approach   The indirect approach that Zulawski and Wicklander (1993) call the *introductory approach* is less likely than the factual approach to cause the suspect to become defensive.

Rather than the suspect talking the majority of the time, the interviewer politely dominates by beginning with a description of the interviewer's responsibilities, background training, and experience. The interviewer then convincingly shifts to the description of the process of an in-depth investigation and the scientific analysis of evidence.

If the interviewer observes that the suspect is listening, then a shift directly to the suspect is made. Similar to the Reid Nine Step method, a face-saving justification is presented and quickly followed with an indirect question about some aspect of the crime. The longer the suspect listens to the interviewer, the higher the probability of a confession, or at least admission.

> "Brian, my job is to protect citizens like you. People want to feel safe in their homes. I've been a cop for twelve years, investigating property crimes for the past six years. We have a lot of break-ins in our town with most break-ins being into citizens' cars. I can't tell you how often we tell people that they are just asking their cars to be broken into when they leave them unlocked or with CD players or GPS units out in the open." (The interviewer makes eye contact and pauses slightly, but not long enough to allow the suspect to deny or say anything. The interviewer observes any changes in the suspect's nonverbal cues.)
>
> "Brian, you probably are not familiar with what goes into a thorough investigation. We use a number of different strategies to obtain information. Lucky for us, more apartment complexes are placing surveillance cameras in parking lots. These cameras are really good. They are able to take clear, wide-angle pictures even in the middle of the night. And no matter how careful, nobody can be sure who is looking out their apartment windows. We knocked on the doors of all of the apartments to see who saw property taken out of the BMW at around 11 P.M. Although the owner was pretty dumb to leave valuable property in his car, he was smart enough to park under some bright lights, which helped with identification. He also had the serial numbers for all of his property so we were able to check it against all of our recovered property. You might not know, but we have an automated system with the pawn shops. They input their pawn tickets everyday so we can match serial numbers.

"Oh yes, we also have informants who give us information from the street. All people have friends and associates, who under the right condition, will provide information. It is only after we have talked to all witnesses and analyzed all of the evidence that we talk to people potentially involved.

"Brian, you have so many pressures on you, and here you were confronted with a temptation. Here was this really nice car that somebody was stupid enough to leave unlocked. You can just be walking down the road, minding your own business, not planning to do anything, and suddenly here is this really sweet car with its doors unlocked. It's like somebody announcing—here I am, take me. These things happen all of the time, and that is what happened here.

"Here is some rich guy driving a really nice car who doesn't even care enough to lock it. You figure he probably doesn't know what it's like to be on the street without a roof over his head. Instead he has it all. It makes us do things we probably wouldn't do normally. It's not like you put a gun to the guy's head and forced him to give you the car. All you had to do was open the door.

"Did you see the driver leave the car or were you just walking by and saw the doors unlocked?"

Reid Nine Steps of Interrogation

J. E. Reid's Nine Steps of Interrogation has widespread popularity among investigators, but it is also controversial (Gudjonsson, 1993). Inbau and colleagues (2005) introduce the steps with a statement that the techniques' goal is to persuade a suspect to tell the truth. While they assert that none of the steps are apt to make an innocent person confess, the authors also caution that the interviewer should be certain of the suspect's guilt. The authors also maintain that the techniques are legal and ethical. North Carolina Supreme Court Justice Exum stated in *State v. Jackson*:

> Many of the practices criticized by the Court were drawn from Inbau and Reid. . . . I think it is unfortunate that the trial court and the majority place such reliance on this book. . . . The truth is that the Supreme Court in *Miranda* was critical of some of the methods suggested by Inbau and Reid, 384 U.S. at 448-56, 86 S. Ct at 1614-1618, and Inbau and Reid are equally critical in their latest version of the *Miranda* decisions (p. 34).

Reid's steps begin with a direct confrontation. A quick evaluation of the suspect's response is obtained. Do they look down and say nothing or cross arms, lean back, and deny the accusation? The interviewer then explains the importance of telling the truth.

> "We have completed an extensive investigation that includes the analysis of physical evidence and the interviews of a number of people. You are the only one we cannot eliminate from suspicion. We have the how, when, what, where. The only unanswered question is why. We know what your buddies say, but we want to hear from you."

The interviewer then introduces theme development. If the crime is property related, the justification for the offense is likely to be impulse, peer pressure, or financial problems. Taking something on impulse addresses temptation. The crime was like a ripe fruit, there for the picking. Peer pressure often allows the interviewer to introduce what the suspect's associates might be saying.

> "It wasn't your idea. It was a group idea. I've talked to Pete and Steve, and they agree that it wasn't their idea. Guess whose idea they said it was. I think you were pressured by them, but they are saying they were pressured by you."

As first described in Chapter 1, Sykes and Matza (1958, as cited in Vito, Maahs, & Holmes, 2007) introduced the Techniques of Neutralization. Interviewers must be aware of the five modes of denial: denial of responsibility, denial of injury, denial of a victim, condemnation of the condemner, and appeals to higher loyalties. Suspects who have a drug or alcohol dependency may blame the influence of the substance or amnesia: "The alcohol made me do it," or "I was crazy drunk then passed out. I don't know what I did."

They may deny the victim was hurt or a victim even existed, such as in the case of larceny from a business: "I worked my fingers to the bone for that company. It got its pound of flesh from me. All I did was take some of what they owed me."

Forms of revenge neutralize acts against former employers, lovers, or teachers. Doctors or other professionals who become involved in certain forms of fraud may plead that their patients or clients could only be helped if they cheated the government or insurance companies.

Social acceptance provides acceptance for anything done by enough people, especially in a crowd: "You aren't the first person who has ever done this sort of thing. I've been an investigator long enough to see it every day. I'm not saying that what you did was right, but a lot of people would have done the same thing."

The theme reinforces the suspect's own rationalization, so it is relatively easy to overcome the suspect's denials. If the suspect is listening and appears to be deliberating, it is an indication of involvement and also investment in the theme being developed.

Although Inbau and associates (2005) deny any ethical issues surrounding the use of theory development, there are risks involved. The suspect's attorney may use the rationalization as mitigating circumstances if the case goes to court. Also, reducing the suspect's sense of criminal responsibility can be detrimental for future rehabilitation.

Blaming the victim is particularly dangerous. In sex offense cases, blaming the victim for acting or dressing a certain way may decrease possibilities of rehabilitation of the suspect. Sex offenders have high recidivism rates, so the interviewer may be obtaining short-term results that impede long-term changes. At a minimum, there are ethical issues.

In the third step, the interrogator continues to develop the theme. Denials should be anticipated, but evaluated. According to Inbau and associates (2005), denials of an innocent person are direct and forceful and will drown out the interrogator's accusation. Weak denials by the suspect should be discouraged: "Let me finish what I am saying."

The interrogator should then anticipate that the suspect will offer reasons why he or she could not have committed the offense. Often the suspect will say the accusation is wrong but then is unable to present evidence of innocence or will attempt to provide distractions from the main crime. The suspect then will begin to withdraw mentally from the interview and try to tune out the interrogator. To retain the suspect's full attention, the interrogator moves closer, maintaining eye contact. The interrogator needs to say something to regain the suspect's attention: "We all make mistakes, don't we?"

The last four steps of the Reid method continue the pressure on the suspect to tell the truth, continue repeating and developing the core theme, and begin to ask questions that help seal the admission and confessions: "I know you are sorry about this, aren't you, Brian? You are a good guy. You would have never taken that car if the keys had not been in the ignition and your friends hadn't been nagging you on."

EXERCISE 10.1

Theme Development

Activity: Students will test one of the primary procedures of Reid's Interrogation Method by developing themes to explain the suspect's alleged reasons for committing the crime.

Purpose: To allow students the opportunity to test one of the main premises of the Reid Interrogation Method.

1. List some of the themes that could possibly be developed to help Warren save face, rationalize, and confess: "Warren, the results of our investigation clearly indicate that your car hit Michael." (Hit and run offense)
2. Respond to the following: As you are outlining the theme, Warren starts shaking his head and says, "I need to tell you. You have it all wrong."
3. Warren gives a big sigh, slumps in his chair, and looks at the floor. You decide it is time to ask an alternative question. What is an alternative question that focuses on the reason Warren committed a hit and run?

Additional Interrogative Techniques

Questions often need to be asked in a nonthreatening way, as discussed in Chapter 4. This method is particularly helpful when working with an experienced offender using the indirect approach.

Implication Questions   Implying evidence frees the interrogator from revealing what has actually been found and analyzed but helps gauge the suspect's level of certainty.

> "Is there any reason that your fingerprints were found on the driver's car door? Is there any reason that a witness would say that they saw you at the Shell station immediately across from the apartment right before the break-in of the BMW?"

If the suspect immediately denies that his or her prints could be on the door or his or her presence at the station, the interrogator can state that he or she would prefer to discuss now if there was a reason for the suspect to be present in that store or around that apartment or to have left fingerprints on the door.

If the suspect believes that he or she might have been observed or implicated by an informant, he or she will have to decide whether to admit to being in the area but not admit to any involvement in the crime. Any delay as the suspect considers options and then changes the alibi probably points to deceptive behavior.

Trap Questions   As mentioned earlier, alibis are usually narrowly developed with few details that the truth would have. Once the suspect believes that he or she has a believable alibi, the interrogator asks a question that will get details from suspect.

> "I'm not saying that you were involved. In fact, you said you had never seen the victim before. Did you know that with today's DNA technology, we have experts who can even lift fingerprints off of human skin? Is it possible that we will find some evidence that you were near the parking lot where the victim was assaulted? We are still interviewing people who live there."

The interrogator sets a trap, and the suspect has a dilemma. There is a chance that there is physical evidence or a witness. If this is a sexual assault, the suspect has to decide whether to consider consent as a defense or remain with the alibi of not knowing the victim. Small cracks start around a minor admission. The alibi is no longer airtight, and because it is not, people will think the worst of the suspect. The interrogator states that "It is better to own up to what can be proven and give your side of the story."

Defeated Protests   Hess (1997) builds on Inbau and associates' handling of denial (2005). He suggests that when suspects resort to reasons it could not be them, it is constructive to agree with them rather than get in a debate. These reasons usually do not directly deny the actual crime: "How could you think it was me? I am on probation and am not supposed to be out after 6 P.M.," or "I love her. How could you possibly think I would hurt her?"

Yes, a condition of probation was for the suspect to remain at home after 6 P.M., and yes, the suspect loved the victim. The interviewer can use the agreement to further the developing justification–argument.

> "I know you loved her and that is why you were hurt and outraged when she left you. All you wanted was for her to come back to you, but in so many words and actions, she told you she was through with you. And then you found her with another man."

Ending the Interrogation

The interrogator should end the interrogation as any interview—professionally and courteously even if there is not a confession or even admissions. The interrogator should be sure to give a business card with contact information and encourage the suspect to call if he or she remembers any other significant details.

Once the interrogator concludes that the suspect is not going to make any admissions, a return to an interviewing stance is important. The questions return to a more open-ended approach, allowing the suspect to talk.

Sometimes, it is helpful to offer a soda or cup of coffee and just sit and chat. Changing the mood might change the suspect's demeanor, allowing him or her to relax and talk about events surrounding the offense. The suspect might brag about his or her activities and associates.

The interrogator should remain alert. The suspect may add to the deception while trying to cover tracks. It also allows the suspect to release some pent up frustrations of being interrogated.

The interrogator should not apologize for confronting the suspect. As the interrogation returns to an interview, the interrogator can politely note that the interview is over and thank the interviewee for his or her time. The interviewer should let the suspect know the investigation will remain open and ongoing.

It is important to leave the door open by ending on a positive note. Before ending, the interrogator should ensure rapport is as well established at the end as at the beginning. Future-oriented questions will leave the suspect with a positive sense of worth also.

The interrogator wants the interviewee to agree to talk to him or her again in the future. After the suspect thinks about the evidence and the rationalization provided by the interviewer, he or she may call the interviewer back.

EXERCISE 10.2

Discussion of Different Scenarios

Activity: To discuss different scenarios with the students.

Purpose: Interrogations are serious actions that should be conducted by professionals. The students will discuss the area rather than practicing specific skills. Use the following discussion questions:

1. A city council member, who is in her first term of office, is arrested for allegedly taking bribes. What interrogative approach would be the most appropriate to use if the person is emotional? What if the person is assertive and calculating?
2. An individual has been arrested in an undercover drug probe. You want to get information from the individual about where his source

of drugs originated, the quantity of drugs involved, and the names of others involved.
   a. What strategies would you use to persuade an emotional subject for whom this is his or her first offense?
   b. What strategies would you use to persuade the subject who is experienced?
   c. Plan an approach for persuading subjects.
3. You are investigating a homicide. After interviewing several neighbors, you believe it was drug deal encroachment in which the victim was trying to move in on another dealer. The neighbors (those who are willing to talk) state that they saw the victim talking to three white males in a blue new model Ford Mustang. The neighbors said that the victim started walking away when they heard a shot and saw the victim fall. They could not identify any of the males. The dealer, who lives across the street from where the victim was found shot, has been cooperative so far, but denies any knowledge of what happened to the victim. The dealer's associate, Jim Brown, said the dealer was with him. You are aware that the dealer owns a Glock semi-automatic. Information that has not been released to the public is that the victim was shot with a round that could have been shot from a Glock. How would you approach the dealer suspect?
4. You are a private investigator. A prominent doctor in town phones you about a woman who is blackmailing him. The doctor is married with children but admits that he was given as a birthday present this woman for a one-night escort. After his birthday, he says he never saw her again but received in his business mail a picture from the birthday night. He wants you to talk to the woman.
   a. What information do you want from the doctor first?
   b. What approach would you use in talking to the woman?
5. You want to have an individual agree to a follow-up interview. How would you phrase the ending?

Conclusion

The courts state that interrogation is not only questioning, but also words or actions by the police that the police know are likely to extract an incriminating response. The interrogation of suspects should be one of the last activities of an investigation with all of the evidence and interviews of victims and witnesses pointing at one specific suspect(s). With more than 6,000 false confessions occurring each year in the United States, innocent lives could be ruined if careful investigation and interviews are not carried out (Sear & Williamson, 1999).

The interrogator should be guided by the Fourth, Fifth, Sixth, and Fourteenth Amendments of the U.S. Constitution. Any statements obtained as a direct result of an illegal seizure (arrest), without probable cause, is inadmissible in court because it was obtained in violation of the Fourth Amendment. Based on the Fifth Amendment, no person shall be compelled in any criminal case to be a witness against himself or herself. According to the Sixth Amendment, once a suspect has been formally charged, he or she can be questioned further only if the prosecution can prove that the suspect volunteered to talk and knowingly relinquished right to counsel. The Fourteenth Amendment outlines due process, the voluntariness and lack of coercion in the questioning of the suspect.

When suspects are interrogated, they have to make some decisions: (1) whether to speak or remain silent, invoking their rights for an attorney; (2) whether to make self-incriminating statements; (3) whether to tell the truth or not; (4) if deciding to tell the truth, whether to tell the entire truth or only part of the truth; and (5) how to answer questions. So, it is the interrogator's goal to decrease the suspect's perception of the consequences of confessing while at the same time increasing the suspect's internal anxiety for lying. The suspect's expectancy will change if he or she can be led to internalize a purpose, or a justification, to tell the truth.

There are a variety of different interrogative approaches, all of which have their strengths and weaknesses and work under certain circumstances. Two approaches are the factual and indirect approaches. Probably one of the approaches best known is the Reid Nine Steps of Interrogation. It is important to be aware of the strengths, weaknesses, and issues surrounding each of these approaches, and interrogating in general.

It is important to leave the door open by ending on a positive note. Before ending, the interrogator should ensure rapport is as well-established at the end as at the beginning. Future-oriented questions will leave the suspect with a positive sense of worth also.

ALLEN'S WORLD

As defined by Dr. Lord I rarely do interrogations. Most often, the suspect is my client and, therefore, voluntarily willing to give me information to help in the defense. However, I must be aware of the same issues Dr. Lord discusses: lying, deception, half truths, and rationalizations.

A constant reminder: Since I am a private investigator, most of the guidelines established by the various amendments to the Constitution, the courts, and public agencies do not apply to me.

For example, I do not recall ever reading suspects or witnesses their *Miranda* rights.

I rarely ask my client if he or she is guilty. I am looking for information to help in the defense.

I tell the client, "Don't lie to me or you will never see me again." I stress that at the moment, and for the term of my investigation, I am God in their eyes, not their attorney, who will not do anything until I finish my investigation, and not the prosecutor, who is in no hurry to move the case along.

Do they have an alibi? Get the details: names, dates, times, places. I get the client to call people who support the client's alibi and tell them to expect to hear from me and to cooperate by answering my questions. In no instance is the client to "plant" information with the people he or she thinks can help.

I do a background check on the folks willing to help. I ask about their eyesight, hearing, medications, criminal history, and drug use. I need to know if they can be impeached on the witness stand.

For some of the crimes such as sexual assaults, the clients may say the accuser was a willing participant. I must probe for the details that might corroborate the willingness.

At the end of most interviews, I will turn off the recorder and just chat with the subject. Usually, they believe the interview is over. Most often, that is when I get the most useful information.

I want to confirm a serious point that Dr. Lord makes. Innocent people can, and do, plead guilty to crimes they did not commit. It is hard to believe, but most attorneys will say the most difficult case to defend is an innocent defendant. Often, the attorney is faced with trying to prove a negative, so a reduced sentence in a plea bargain seems like a sweetheart deal when weighing the sentence based on a guilty verdict.

Arnold, an elderly gentleman with a dignified manner, was accused of numerous counts of molesting four young children.

I interviewed Arnold at length, toured the alleged crime scenes, and interviewed his wife. I interviewed some of the parents of the children involved.

I was convinced that Arnold did not commit the molestations, but it is difficult to cross-examine children without gaining sympathy from jurors.

The prosecutor offered Arnold a deal. Plead to some misdemeanors, and he would receive a maximum sentence of 24 months and be done with it. The alternative? Go to prison for the rest of his life. Arnold reluctantly took the plea, served his sentence, and is back with his wife.

Crisis Intervention and Negotiation

OBJECTIVES

Upon completion of this chapter, the student will be able to

1. Understand the general concepts of crisis negotiation
2. Provide a general overview of barricaded and hostage negotiation
3. Describe characteristics of an effective negotiator
4. Describe the categories of subjects in barricade situations
5. Understand the dynamics between the negotiator and the subject
6. Provide a general overview of the strategies used in crisis negotiations
7. Describe an overview of the approaches developed in barricaded–hostage situations

Introduction

Crisis negotiation is a specialized area of law enforcement that depends heavily on communications. Only a handful of police and police-affiliated interviewers are involved in negotiation. Many of the same skills are needed for negotiation as are needed for investigative interviewing. Complete, valid, and relevant information must be gathered from people associated with the barricaded subject or hostage taker. Rapport must be developed between the subject and the interviewer-turned-negotiator.

Special response teams have existed since Biblical times. The Old Testament's description of Abram's use of 318 men to rescue Lot may have been the description of the first Specialized Weapons and Tactics (SWAT) team. The kings of Sodom and Gomorrah fought against the kings of Elam, Tidal, Shinar, and Ellarsar and were defeated. Lot, Abram's brother, who lived in Sodom, was captured. Abram with his 318 men surprised the kings of Elam, Tidal, Shinar, and Ellarsar at night and chased them into Hobah. Abram was able to bring Lot and all of his property back to Sodom (Genesis 14: 14–17 Revised Standard Version).

SWAT teams continue to be an important component of hostage or barricaded incidents; however, negotiations usually are the first step in preventing loss of life. When negotiations fail, SWAT team intervention becomes mandatory.

The world was horrified when, during the 1972 Olympics, 11 Israeli athletes were seized by Black September, terrorists affiliated with the Palestinian Fatah movement. SWAT teams, snipers, and negotiators descended on the Olympic Village. The terrorists' demands included the release of 200 Palestinian prisoners, who were being detained by Israel. When Israeli negotiators offered alternatives, such as free passage to China for the hostage takers, money, or trade for personnel other than the Palestinian prisoners, the terrorists thought the Israelis were stalling and threatened to kill two of the athletes immediately. Sixteen hours after the athletes were seized, 22 people were dead—one police officer, 10 terrorists, and the 11 Israeli hostages. Afterward, police from all over the world examined the negotiation and tactical techniques used by the Israeli and German SWAT teams who attempted to negotiate and then shoot their way out of the crisis (McMains and Mullins, 2006).

Hostage negotiation was born and quickly became tested. In 1973, New York police had the chance to try out the newly developed basic principles in a hostage situation. Local police responded to a silent alarm, possible robbery in progress, at John & Al's Sporting Good Store. The officers were met with gunfire. Eight hostages were held. Two officers were injured. The newly trained tactical unit responded. Its primary concern was containment of the hostage takers, control of personnel and resources, and communication with the hostage takers. Forty-seven hours later all of the hostages were safe. Four perpetrators were in custody, and no more officers had been injured (McMains and Mullins, 2006). The situation was concluded successfully without any loss of lives.

How did the police successfully negotiate the release of the hostages?

Harvey Schlossburg and Frank Boltz developed three key principles (Schlossberg, 1979, as cited in McMains and Mullins, 2006): (1) the hostage taker must be contained and negotiation implemented; (2) the hostage taker's motivation and personality must be understood; and (3) the incident must be slowed down so that time can work for the negotiators. Negotiation is the safest approach with the goal of no loss of life.

In order to understand crisis intervention and crisis negotiation, a number of concepts need to be understood and differentiated. The following are definitions for some of those concepts.

Crises are circumstances perceived as catastrophic. Individuals are faced with serious consequences and feel overwhelmed and unable to cope because they do not have the necessary resources, including past problem-solving tools. Some of these individuals may barricade themselves and threaten suicide or homicide. They feel helpless and hopeless.

Crisis intervention takes place in highly emotional situations such as interviews with victims immediately after the crime has occurred, or with first-time offenders facing probation or an active prison sentence. The subjects are in crisis because they are facing situations in which they probably do not have pre-existing coping mechanisms to deal with the situations.

Crisis negotiation is the general term that encompasses any situation that police confront in which a subject or subjects involve the police in a life-threatening incident. The threatened lives may be those of the subject, hostages, bystanders, or the police. Crisis negotiation then is the interaction between negotiators and subjects (hostage takers if they are holding hostages) in a tense environment. The goal of the negotiation is settlement, paying attention to meeting legitimate interests of both sides, but keeping the community interests—and the safety of any hostages—in the forefront.

A *barricaded situation* is any incident in which one or more armed or potentially armed persons fortify themselves within a protected location (building, house, vehicle, bridge) with or without hostages and refuse to surrender to the police. These situations often begin as domestic disputes, suicide attempts, or crimes.

Hostages are people who are held against their will as a means to secure specific terms by their captors. These hostages are not in the situation voluntarily, but rather, because their lives are threatened. Often they are in the wrong place at the wrong time. They will become traumatized by the incident, feeling powerless and vulnerable. Although the hostages are human, they are usually seen by their captors as little more than commodities with value. It is important to personalize the victims, but also not to increase their value and consequently the hostage taker's power.

All criminal justice personnel interact with people in crisis. When in crisis, individuals feel as if their lives are out of control. Because their emotions cloud their ability to reason, their ability to problem solve is reduced. Crisis intervention's goal is to return the individuals to their baseline level of functioning before their crises.

Types of Subjects and Situations

As mentioned, crises can be precipitated by any number of different situations. The four most common categories are trapped perpetrators, couples involved in domestic disputes, suicidal individuals, and individuals with mental illness. Lord and Gigante's study (2004) records almost an equal division among suicide attempts, domestic disputes, mental illness, and crime-related situations, while Amendola, Leaming, and Martin's study (1996) primarily reveals suicide attempts or bungled criminal cases.

Each of the categories has specific characteristics that must be considered when negotiators interact with them.

Trapped offenders are subjects caught by police during commission of crimes. Their only motive is to escape. They do not intend to be apprehended. They will take any action to avoid prison time. Consequently, they are unpredictable, volatile, and dangerous.

Individuals with mental disorders or those experiencing acute emotional turmoil emerging from personal problems or family disputes normally do not take hostages. If they do, the hostages usually are family members. The FBI analyzed 245 incidents of barricade–hostage taking and concluded that more than half (145) of the perpetrators suffered from a mental disorder or were experiencing emotional turmoil from personal or family problems

(Slatkin, 2002). The disorders most likely to result in violence are mood or cognitive disorders such as depression, bipolar disorder, and paranoid schizophrenia (Soskis & VanZandt, 1986). These individuals usually are loners without prior criminal histories.

Many of the mentally ill individuals have a treatment history, and often it is possible, almost mandatory, that the negotiator contact the individual's therapist. The risk in contacting the therapist is that the subject may be angry or disappointed with results of the therapy, so direct intervention by the therapist may increase the subject's level of hostility and agitation. On the other hand, information supplied by the therapist might provide the negotiator a better understanding of the issues and a better chance of negotiating a peaceful surrender. Negotiators often find people with a mental disorder are acting out because they have not taken their medication. Self-medication with controlled illegal drugs or alcohol also is a common failure.

Fuselier (1986) argues that approaches negotiators take should differ for dealing with the mentally ill, suicidal folks, couples involved in domestic disputes, and trapped perpetrators.

For those individuals who are mentally ill, the negotiators should work hard to establish rapport rather than attempt to address their delusional problems because the subjects will not be able to interact rationally. Individuals involved in a serious domestic dispute also have emotional issues, but if the negotiators communicate support and understanding, they might be able to move the subjects to a more cooperative mode. Those individuals involved in criminal activity can be addressed more directly. Negotiators can address factual issues and focus on exchange of items for hostages.

"Suicide by cop" is a term used for those individuals who, when confronted by law enforcement officers, either verbalize their desire to be killed by the officers or make life-threatening gestures, such as pointing a weapon at them. They have been typed similarly as other barricaded subjects; the critical difference is their desire to die at the hands of police officers (Lord & Gigante, 2004).

Another form of barricaded and hostage incidents for which police are beginning to train is *terrorism*—defined as the use or threat of violence to achieve social, political, or religious aims. Acts of terrorism are carefully planned. The terrorists' religious and/or political beliefs may be strong enough that they are willing or even anxious to kill and die for a cause. It would be hard, for example, to talk a suicide bomber out of pulling the cord.

Characteristics of Negotiators

It takes special officers to negotiate. Negotiators must be able to establish rapport, reduce emotionality, and convince subjects to yield to the negotiators' demands that include the release of hostages and a peaceful surrender.

Verbal skills are the obvious needs. Officers who are good listeners, have demonstrated excellent interviewing and persuasive skills, and can establish credibility easily should be selected as negotiators. The negotiator must be able to communicate concern for the subject's needs and interests. The negotiator needs to be emotionally and professionally mature with the ability to function well under stress. Other important characteristics are patience and flexibility. Time is required to defuse emotions and to help the subject rethink his or her position and demands.

A positive self-image helps the negotiator cope adequately with uncertainty and the willingness to accept responsibility with limited or no authority. Most police agencies include periodic psychological evaluations of their negotiators and SWAT members.

All of these attributes come with experience. The FBI recommends at least 5 years of experience before someone becomes a negotiator. Two negotiators for every situation are recommended for psychological support (Slatkin, 2002). Under adverse and stressful circumstances, these negotiators must develop trusting, helpful relationships that lead to resolution with a minimum of injury or loss of life. These negotiators must let the subject know they can act as a conduit. In other words, they have a link to, but not power over, their superiors. The negotiators foster a belief that they have similar concerns as the subject in regards to resolving the situation. Although they may have different views on what "turning out well" means, they agree that a violent outcome would be negative for all.

Dynamics of Crisis Situations

Criminal justice personnel in a variety of fields besides police come in contact with people who are having difficulty coping during a crisis, beginning a slippery-slope cycle. Poor coping skills are often what lead people to criminal activity, and poor coping usually leads to more frequent crisis experiences. For example, when people lose their jobs, which creates financial hardships and lack of self-esteem, they have a number of solutions they can consider. They may do one of the following:

1. Look for similar jobs for which they have experience
2. Return to school for additional training in their current area or training in a new area
3. Move to a new location with prospective jobs
4. Do nothing except watch television and complain about how life is not fair.
5. Drown their problems in alcohol or drugs

Increased use of alcohol and drugs can lead to further crises. People with poor coping skills that include excessive use of alcohol and drugs may end up losing everything—their families and their homes.

When in crisis, individuals feel as if their lives are out of control; their emotions are high, and reasoning is low. Their ability to problem solve is reduced. They may end up in a confrontation with authority, whether it is the police, their probation officers, or their social workers. They are trying to regain control of their lives.

"I don't know what I am going to do. I can't believe she left me for that SOB. I may not be the greatest guy in the world, but we have been together for five years. We have two babies. How can she leave me for him? He's got to go. I know that for sure. I might just kill myself then she would be left alone. That bitch would know what it feels like to be all alone. Let her try to raise our babies by herself. Let her be all alone like she left me. What's the point of living? She left me, took our babies, and now she is with him. I can't stand the idea of his hands on her or around our kids."

The focus of crisis intervention is to return the individual to the level of functioning before the crisis. Intervention is most effective if implemented as close to the onset of the crisis as possible. Crisis intervention can be used by nonmental health professionals as long as they have received additional training (McMains & Mullins, 2006).

Dynamics of Hostage Situations

According to Crelinsten and Szabo (1979), there is a triangular aspect of hostage taking; the hostage is the means by which the hostage taker gains something from a third party. The passive victim (the actual hostage) is a means to an end, an involuntary intermediary in an exchange between the offender and active victim (the individual who can meet the hostage taker's demands). The relation between the two victims is critical. If the active victim feels no great concern about the passive victim, then he or she is unlikely to yield to the demand and avert the threat.

The active victim can choose to act alone or to involve other parties, such as superiors, police, friends, or relatives, who then become part of the process. A feedback mechanism is set up whereby the response of the primary victim and those whom he or she calls into the case feedback to the offender who then alters either the threat or the demand or both.

Whether or not a hostage is involved is a critical factor in how much power and control the subject has and, therefore, greatly influences the officers' response. Although the officers' goal is to bring a peaceful resolution to the incident, if there is no hostage, there are a number of options available to officers in addition to negotiation. Turning off the electricity, using chemical agents, and even resorting to tactical entry and assault are considerations.

Donohue, Ramesh, and Borchgrevink (1991) note that the goal of negotiation is to move the hostage taker off a coercive stance to a cooperative stance. In a coercive stance, the subject will reject any need to develop a relationship with negotiators. The subject uses his or her power over the hostage to attempt to force demands.

Donohue and his colleagues studied the dynamics of hostage situations based on the three categories of subjects: criminal offenders, mentally disturbed individuals, and domestic disputants. The researchers then examined the common form of negotiation behaviors dictated for each category.

Criminal offenders begin in a competitive mode that escalates, but then drops into more normative bargaining once the initial crisis phase has diminished. Toward the end of the incident, there are brief periods of competition as the offender feels some degree of anxiety in releasing hostages and/or surrendering.

Mentally ill individuals are less predictable. They move quickly into the cooperative phase, but then return to a competitive mode and remain entrenched until the situation is resolved through either surrender or suicide. Individuals who are mentally ill may appear cooperative to deceive the negotiators into thinking they will do whatever the negotiators want.

Hostage takers use their hostages as their pawns to coerce or communicate with the negotiators. The offender (1) communicates threats to the hostages; (2) the hostages make pleas to the offender and the negotiators; and (3) the offender hopes these pleas will help reinforce direct messages he or she is sending to the negotiators. If the offender is mentally or emotionally ill, the triadic dynamics may not apply. If the subject is paranoid with delusions, the police may not exist, or do not have a meaningful connection to principals involved in the subject's delusions. For example, the subject may demand that international politicians bring immediate peace to the world, or aliens, who are taking over the world, should be forced to leave. Subjects with mental illness are particularly difficult to develop rapport with because they may be deficient in the ability to develop intimate, trusting human relationships.

Donohue and his colleagues (1991) found that domestic hostage takers take longer to reach the cooperative mode than the criminal offender. Similar to the criminal subject, as terms of surrender are discussed, there is a second crisis period that will put the domestic disputant back into a competitive mode.

Dynamic Factors   Fuselier, VanZandt, and Lanceley (1991) group antecedent events that influence the subject into the three areas of financial, family, and social pressures. In addition, many of the perpetrators come from a background in which male dominance is encouraged, so when their perceived ability to be a man is threatened, negotiation may become more difficult.

Fuselier and colleagues also note that these situations can be planned or spontaneous. Planned hostage-taking goals include freeing a prisoner, forcing political or social change, or publicizing a cause or perceived wrong. Spontaneous incidents can be particularly dangerous because these subjects may be willing to die for their cause.

Most barricaded situations are not planned. They begin as a crime that suddenly goes wrong—a fight between significant others that escalates when the police arrive or one disputant tries to leave, or an emotionally unstable person who becomes delusional.

There are a number of situational factors that play into the interaction between the offender and any hostages he or she might have, as well as between the subject and the negotiators. The situations involve not only physical factors, such as weather, media, bystanders, and the structure in which the subject is barricaded, but also the physical and emotional state of the offender.

Time is critical, and the duration of the incident is of particular importance. Reducing tension is an important early objective for negotiators. Kupperman and Trent (1970) claim that the first 3 to 4 hours are most dangerous, noting if nobody is killed by then, there is a good chance that everybody will remain safe. Head (1987), however, concluded that the first half hour is the most critical because the offender is most likely to be unclear of his or her plans.

McMains and Mullis (2006) note that time increases basic human needs, and the negotiator can be placed in a position to fill those needs. As the offender gets hungry, the negotiator can manipulate the appeasing of his or her appetite to gain trust and to reduce anxiety. The actual time in which these situations occur compares favorably with other forms of violent crimes. The evening hours in which significant others are likely to spend time together and perhaps imbibe in a few drinks appear to be the most frequent. There are certain times of the year that are emotionally more volatile. Similarly to suicide rates, early spring, late December around Christmas and the New Year, and the dog days of summer seem to be particularly prevalent.

Negotiators are able to use extreme temperatures, especially the cold, to help them influence the process. It is not uncommon to cut off the electricity when there appears to be an impasse in the negotiation process.

Negotiators often view the media as a constraint, although some negotiators have learned how to manipulate the media to their advantage. Media often will give too much information to the subject who may be actually watching himself or herself on television. Media also will frequently draw a crowd that then must be controlled. Media can interfere physically with the police actions. On the other hand, negotiators have learned to actually use the media as their voice to the subject. If the negotiators believe the subject may be watching television, they will be interviewed and speak about the subject in as positive terms as is possible to help build rapport. For example, negotiators may note that the subject is dealing with a number of problems but seems to want to do the right thing by his or her family.

Bystanders always are seen as an impediment to the officers' actions, especially if they include family members. If bystanders are an audience for the offender who is of a grandiose nature, these gawkers provide more ammunition for the offender who wants to "leave in a blaze of glory." A new development is the use of cell phones to communicate with friends and family members. In one case, the police began to confiscate family members' cell phones when they realized the family was telling the subject where the tactical officers were stationed (Lord & Gigante, 2004). This new development is in addition to the normal family advice, pleas, and retorts. In all cases, bystanders are an additional responsibility for the officers. Regardless of how they are behaving, bystanders must be protected at all costs.

The influence of alcohol and drugs is an important factor to be considered when decisions are made tactically. In most cases, it is important to keep the subject talking so he or she might sober up. In other cases, it has been found useful to let the subjects drink until they became too tired and drunk to continue. For example, in the early 1990s, in a small town in North Carolina, police responded to a hostage situation at a bar. The situation began with a female drinking in a bar until the bartender refused to serve her any more drinks. The woman staggered out to her car and returned with a gun, holding the bartender and several patrons hostage. A police negotiator called the bar, and the woman picked up the phone. She screamed at the negotiator, stating that the bartender had refused to sell her drinks, and her children had been taken away from her. All the while, the woman continued drinking. As the night wore on, she became tired. She kept hanging up the phone. The negotiator kept phoning her, keeping her awake. Finally, she let the bartender and the customers go and then came out. Although it might appear risky to let an individual continue to drink, in this situation any alternative tactical strategies could have gotten the hostages hurt or killed (Lord & Gigante, 2004).

Stockholm Syndrome On August 23, 1973, an armed subject entered a bank in the Norrmalmstorg area of Stockholm, Sweden. He held four bank employees hostage for 6 days, demanding the release of a friend from prison and $750,000. The bank and authorities refused to honor the demands. During the siege, the hostages became emotionally attached to the hostage taker and feared the police. The phenomenon of the switch in allegiance and emotions of the hostages became known as "Stockholm Syndrome" (DeFabrique, Romano, Vecchi, & VanHasselt, 2007).

A highly publicized example of the Stockholm Syndrome occurred shortly after the 1973 hostage taking in Sweden. On February 4, 1974, the Symbionese Liberation Army, aka SLA, kidnapped millionaire heiress Patty Hearst. Hearst was shown involved in an armed robbery planned by the SLA on April 3, 1974. There is a controversial picture of Hearst in the bank, armed with an assault weapon, and appearing to participate on her own. A month later, the SLA robbed Mel's Sporting Goods Store. Hearst, again, was seen as an active participant.

Was Hearst really participating freely? Had she been brainwashed by the SLA? Had she fallen victim to the Stockholm Syndrome, or was she a captive of the SLA and under constant threat? In 1976, a jury decided that Hearst was a willing participant and convicted her of armed robbery, not believing the powerful legal defense team headed by F. Lee Bailey that Hearst was suffering from the Stockholm Syndrome. The judge sentenced Hearst to 7 years in prison. After serving 22 months, President Jimmy Carter commuted her sentence, and on his last day in office, President Bill Clinton gave Hearst a full pardon (Castiglia, 1996).

The Stockholm Syndrome has come to mean that after a number of days of captivity, hostages develop positive feelings toward the hostage taker and negative feelings toward the authorities who are attempting to resolve situations. That defense will be on the legal landscape for years. Psychologically, the Stockholm Syndrome is explained as hostages perceiving themselves endangered by the subject, but also under his or her control. The offender has total control of the hostages' lives. The offender can precipitate a fatal ending through unpredictable action, so any behavior by the offender that appears the least bit caring or nurturing toward the hostages possesses tremendous emotional value and forms the basis for positive identification. The hostage incident places the hostages in situations of powerlessness and danger similar to experiences in childhood. What does a scared child do during a thunderstorm? Attaches to the closest parent and seeks comfort and reassurance.

Recently, there has been some skepticism about the syndrome and its actual existence. In a review of 12 cases that were categorized as involving Stockholm Syndrome, a team of psychiatrists, who published their results in a professional Scandinavian psychiatric journal, found no uniform pattern of features. The necessary time and intensity hypothesized to be required was not present in the 12 cases. McMains and Mullins (2006) conclude that the Stockholm Syndrome does not occur in most, if any, cases of hostage taking. They argue that not only are the incidents too brief and lack the necessary intensity, but also most hostage takers and their hostages already have an emotional or professional relationship; they rarely are strangers.

There are some recognized variations of the Stockholm Syndrome categorized as "trauma-bonding." Trauma-bonding is common among victims of abuse in which the victims attach to the abuser due to the perceived power of the abuser to maximize their chances of survival.

When interacting with hostage takers, negotiators must recognize the strong ties that might develop between the victims and offenders, especially if there was a previous domestic relationship. The victims may minimize the potential violence of the hostage taker and even speak favorably to the police, or later the media, about the hostage taker. As the hostages are released, they need to be isolated and carefully evaluated with the help of psychologists.

On the other hand, the Stockholm Syndrome can be considered a potential survival tool for the victims. Developing a relationship with a hostage taker, who is unknown to them, will help personalize the victims. The offenders might come to view them as humans with feelings, families, and personal lives.

Crisis Intervention

As mentioned earlier, criminal justice personnel must be trained in the management of crisis situations. If the subject is a client of a probation officer, then quite often the officer, through conducting a psychosocial assessment, has much of the information needed for the first step of crisis intervention but may need to expand information about the precipitating event (McMains & Mullins, 2006; Samantrai, 1996; Walsh, 1988).

- Why is the individual in crisis now?
- What are the precipitating events that led up to the crisis?
- Has the individual attempted to deal with the crisis in the past?
- How successful was the coping?
- What resources are available? What resources are needed?

If the individual is confronted by the police or is unknown to the criminal justice personnel, then the interviewer must collect information in much the same way as with the beginning of most interviews. The interviewer must develop rapport and a relationship with the subject–interviewee and identify and reflect the content and feelings of the subject's problems. The interviewer utilizes the characteristics and techniques described in earlier chapters.

> The young man sat in front of the juvenile court counselor with his eyes cast down. His hands were shaking.
>
> "I've never been in trouble before. What are my folks going to say? My girl-friend's family won't ever let me see her again. I've been accepted to the varsity football team."
>
> "It was just a prank. We didn't mean to do anything to hurt anybody. We just threw that old couch over the bridge in fun. Now you are telling me that the car it hit hurt the driver bad."

Slatkin (2002) argues that negotiators should be trained in *therapeutic communication*, especially to be used when subjects have mental or emotional disorders. Heavy emphasis is placed on listening, paraphrasing, reflection of feelings, clarification, empathy, and summarizing.

EXERCISE 11.1

Review of Active Listening Skills

Activity: Applying active listening skills to negotiation examples.

Purpose: To illustrate the importance of active listening skills to negotiation incidents.

    With each response, match a reflection of feeling and content response.

1. "You can't help. Nobody can. I want to be left alone."

   _____

2. "Look, even with the medication that the psychiatrist put me on, I still can't concentrate. That's why I lost my job. The police can't help me concentrate. Now it's just going to happen again. I'm tired of trying."

   _____

3. "I've got a gun. Don't try and come in here."

   _____

4. "I'm at the end of my rope. I can't think anymore. Nothing is going to work."

   _____

5. "I can't live without her. She's running around on me, but I still love her. I'm keeping her here with me until she understands and won't leave me."

   _____

In barricaded situations, the negotiator will still need to collect information from friends and family as well as the subject. The interviewer–negotiator next will work with the subject to brainstorm about resources and all possible solutions to the problem(s). In office situations in which probation officers may be working with clients in crisis, or juvenile court counselors are working with children and their families in crisis, brainstorming solutions can proceed at a leisurely pace, sometimes over more than one session. The key is that the clients know somebody has heard them and is helping them resolve an intolerable situation.

> "I can tell that you are frightened and very concerned about your future and the woman who was driving the car. This is a very serious matter, but it is important that you have never been in trouble before, and your school record substantiates your good conduct."
>
> "Take some deep breaths. Now let's take your fears one at a time and brainstorm some solutions."

Therapeutic communication includes the importance of providing some accurate information when relevant, e.g., "The City has resources in its community development office that could help you deal with that problem."

Also, the importance of reinforcement is later discussed as part of Hammer and Rogan's S.A.F.E. model (2005, as cited in McMains & Mullins, 2006). "You told me that you served time in Iraq with the Marines. I think that says a great deal about your strength of character."

In barricaded situations, especially if hostages are involved, time should be slowed, but resolution is required within a realistic time frame. Finding solutions for each part of the problem that are acceptable to both sides will be essential even if they are not perfect solutions. A subject who has committed an offense and is barricaded cannot expect to go free but may be willing to surrender when he or she realizes the offense is not sufficiently serious to receive active prison time, but rather he or she is more likely to get probation, community service, or a little time in the local jail.

> "Let's work together to find some solutions that will keep everybody safe. I know that is important to you. We don't need to be in a hurry. I want to hear what you think will work, and I may have some ideas especially in the way of resources or services."

Important communication continues with the creation and implementation of a plan. It is important to keep the subject involved so the solutions, planning, and implementation become a partnership. The more the offender believes he or she is contributing to working toward a solution, the more likely he or she will be willing to buy into that solution.

EXERCISE 11.2

Crisis Intervention

Activity: Group discussion about intervening in a client's crisis situation.

Purpose: Apply the crisis intervention steps.

You are a probation officer. One of your clients has missed her last two office visits, so you decide to go to her home. You knock on the door. It takes her awhile to get to the door. When she sees you, she becomes upset and will not let you in. When you look past her, you notice that her house is filthy. Discuss how you would take the client through the crisis intervention steps.

Negotiation Approaches

Hammer and Rogan (1996) developed a model called the communication approach that combines techniques from the basic bargaining negotiation approach built on social exchange theory and the expressive negotiation approach that primarily focuses on the emotional level of the subject.

Based on the social exchange theory, the bargaining negotiation approach is simply the identification of demands and clarification of terms for exchange of resources. According to the social exchange theory, negotiations involve rewards and costs for each party; the negotiation action involves the exchange of some object or commodity in return for other objects or commodities. Overall effective negotiation is the result of rational discourse between the contending parties. Donohue and associates (1991) and Fusilier (1986) would argue that the negotiation with criminal offenders would benefit from this approach as they usually have fairly direct and coherent demands.

In terms of barricaded criminal offenders, the bargaining negotiation approach would have the offender work for everything he or she gets, but get something in return for each concession given. Like a swap meet, "If I give you this, what will you give me in return?" the negotiators would use that extra time to their advantage and not give too much too soon. For example, the offender might get a cigarette for letting a hostage go. Most types of crisis situations law enforcement encounters do not match the requirements of bargaining negotiation. These subjects often are barricaded as the result of mental and/or emotional inability to cope with life's stressors, so their ability to bargain rationally is unlikely.

The expressive negotiation approach focuses on the impact of emotion and relationship and is based on three principals. The first premise is that if there are hostages, they rarely function as a bargaining chip to achieve specific outcomes but rather as expressive acts of the offender to demonstrate his or her ability to control others. The second premise is that the interest of both the offender and negotiator is to prevent the situation from escalating to injury or death. The third premise is that the high emotional level reached during a crisis situation can negatively impact a negotiated outcome. Emotion is such a central element that the likely response is the instinctive "fight or flight" rather than problem solving. Relationship and trust building are viewed as critical factors in resolving crisis incidents.

Hammer and Rogan modified the bargaining and expressive negotiation approaches mentioned above and developed the communication negotiation approach. They concluded that the negotiation process is dynamic and dependent on the interaction between the negotiators and the subjects. Hammer further modified the communication negotiation approach to the S.A.F.E. model (Hammer, 2005, as cited in McMains & Mullins, 2006). In the S.A.F.E. model, Hammer and Rogan identify four important elements that impact the escalation and/or de-escalation of the situation toward or away from increased violence and potential injury: substantive demands, attunement, face, and emotions.

Substantive demands are those instrumental demands made by the offender (I want a cigarette), and the counter-expectations made by the negotiator (we need to know that the hostages are safe).

Attunement is the development of the relationship between the negotiators and the subject. Power and trust are core elements of the relationship between the offender and those in power such as the police. Power concerns dominance and submission. The offender desires dominance and resists submission. Trust revolves around the degree to which the subject and negotiators are willing to accept the idea that a future act by the other will not be detrimental. The negotiator will have to help the subject find benefit in submitting. The subject's emotions must be identified and managed.

Face is the offender's perceptions of his or her own attributes, but also often has a social identity. Saving face is a principal component of the identity concern. There are six types of face message behaviors. Four are the offender's messages, and two are the negotiators.

1. The offender defends own face with self-directed messages designed to protect his or her self-image: "It's not my fault."
2. The offender attacks own face with self-directed messages against his or her self-image: "I have nothing to live for."
3. The offender attempts to restore own face: "I'm not as crazy as you think."
4. The offender attacks the negotiator: "You're jerking me around."
5. The negotiator attempts to restore the offender's face: "You have lots of people who care about you."
6. The negotiator defends, and at the same time, protects the offender's face from future attack or loss: "I think you are really a strong person for how you've handled this situation so far."

Hammer and Rogan (1996) argue that if the negotiator listens to which dimension the offender is communicating, the negotiator will be able to interact effectively with the offender. Hammer and Rogan found that restoring other's face was the primary face behavior used by negotiators, while restoring self's face commonly was used by subjects.

Lord and Gigante (2004) provide a suitable example of the importance of saving face. A juvenile male with a minor criminal and drug record was upset because his girlfriend was in the process of being adopted by a relative who was placing constraints on her life and his ability to see her. The boy broke into the house, hit the father with the butt of a gun, and shot out the television. The family ran out of the house, including the girlfriend. Negotiators came to the scene. The boy began very aggressive, attacking other's face, with statements of "Just kill me. That's what you want anyway. I'll not come out alive." The negotiators worked on restoring other's face by getting him to talk about his feelings for his girlfriend and his need to take care of himself. The boy would catch himself lowering his aggressive response and would once again begin to threaten, or attack other's face. He also wanted to get messages to his girlfriend that could be interpreted as restoring self's face. These messages included that he loved her and was sorry that he had attacked her family. The negotiators in return used messages of defending other's face. These messages were that he had been trying to help his girlfriend and to take control of the situation. The cycle continued for several hours until the boy finally agreed to come out. No shots were fired. No property was destroyed by the negotiators.

EXERCISE 11.3

Saving Face

Activity: Construct responses to face-threatening responses.

Purpose: To help students apply face-threatening and face-saving interactions.

1. "Her father hates me. I always tried to be polite around him, but it wasn't enough. Now he won't let me see her."

2. "I have nothing to live for. The county took my children. They said I was a bad mother."

3. "You don't really care about what happens to me. You are just trying to be nice so I will surrender and you can go home to your warm house and family."

The negotiator must continue to be positive and patient. It will take some time for the offender to move from berating or belittling the negotiator, who symbolizes authority and all that is wrong with the offender's world. When possible, the negotiator should reframe the attack, e.g., "It sounds like you have had some bad past experiences with the police. What can we do so that it doesn't happen again?"

The offender will be concerned with his or her safety, so the negotiator must be able to help the subject work through the consequences of his or her actions. Usually, perpetrators believe their actions are more serious, and the consequences much worse than they really are, so the negotiator can explain and thus minimize them to some extent. Past these two needs, the negotiator can then begin to address the real issues, nourishing the perpetrator's social and self-esteem needs. The negotiator should try to accumulate small positive steps. By first finding areas of common interest and then honoring small requests, the negotiator begins to accumulate "chips" for a major concession.

When the end is in sight, the negotiator should slow down even more, review agreements, and explain what will happen next. Too often, when all appears to be progressing satisfactorily, the hostage taker or barricaded subject gets anxious and foresees the loss of power coming. The continual need to save face is critical, and resolution often is seen as the ultimate face loser. For example, it may be important to remove the media or any crowds from near the area, or the negotiator may help the offender write a victory speech (McMains & Mulllins, 2006). At the conclusion of the negotiations, the offender is more likely to surrender quietly and safely if he or she feels that he or she has a choice, is seen as a person, and can save face.

Criteria Measurements   Positive progress in a negotiation is indicated by less violent content in the offender conversing with the negotiator; the offender talking more with the negotiator; the offender talking at a lower rate, pitch, or volume; and the increased willingness on the part of the offender to discuss personal issues. Positive signs toward the release of hostages include negotiations getting past a deadline set by the offender without incident, a decrease in the number of threats made by the offender, and the absence of violence. On the other hand, negative indicators include the disclosure of information that the offender has recently killed a significant other, especially a child; the offender demands that he or she be killed by officers; the offender refuses to negotiate; the offender sets a deadline for his or her own death; and the offender has a past history of violence.

Always in the background is the decision to proceed tactically. As was learned by the Bureau of Alcohol, Tobacco, and Firearms (ATF) and the FBI in the 1993 Branch Davidians-Waco barricaded incident, negotiations and tactical teams must work together. ATF was in charge of the raid. Three teams of ATF agents, between 70 and 76 agents, were to surround the Branch Davidian compound and secure it before any of the Davidians could get additional weapons. Unfortunately, David Koresh and his followers were warned, and a lethal firefight ensued. It has never been determined who fired the first shots, but when the gun battle ended, four ATF agents and five Davidians were dead. Sixteen ATF agents and four Davidians were wounded (McMains & Mullins, 2006).

At that point, the FBI assumed command and negotiated for 56 days, obtaining the release of any of the children being held hostage. In return,

Koresh was able to publicize his message to the public. Then, a tactical decision was made. An assault took place. The compound went up in flames, whether from the assault or by Koresh lighting the torch has never been decided. A large number of the Davidians died in the conflagration.

Much of the controversy surrounding the negotiations and firefight are described by hostage negotiator Christopher Whitcomb in his book *Cold Zero* (Whitcomb, 2002). Difficult decisions are made surrounding the risk to victims, officers, and subjects. The mission of tactical and negotiation units is to save lives. After the Branch Davidian-Waco incident, the FBI developed a set of criteria to be considered before intervening tactically, ordering SWAT to force its way into the barricaded situation (Noesner, 1999). The tactical and negotiator supervisors should be able to answer the following:

1. Is the action necessary at this moment?
2. If the answer is yes, what has changed to make tactical entry necessary?
3. Is the entry acceptable? Can we document the justification so it will stand up to media and public inspection tomorrow?

Conclusion

Negotiations are a special type of communication process that shares many of the techniques of effective interviewing. While few professionals will become official negotiators, many will help clients, disputants, and the public deal with crisis situations. Following techniques learned as good interviewers and communicators will help facilitate the resolution of crisis, barricaded, and hostage situations.

Individuals in crisis feel overwhelmed and unable to cope because they do not have the necessary resources, including prior problem-solving experience. They feel helpless and hopeless. When in crisis, individuals feel as if their lives are out of control. Because their emotions cloud their ability to reason, their ability to problem solve is reduced. The goal of crisis intervention is returning the individual to their baseline level of functioning before the crisis. Understanding some of the specific dynamics of these situations, along with effective communication skills, will help the interviewer–negotiator effectively intervene.

Crisis negotiation is the general term that encompasses any situation that police confront in which a subject or subjects involves the police in a life-threatening incident. The goal of negotiation is to move the subject off a coercive stance to a cooperative stance. Helping subjects restore face is an essential key to negotiating successfully. At the conclusion of the negotiations, subjects are more likely to surrender quietly and safely if they feel they have a choice, are seen as human beings, and can save face.

ALLEN'S WORLD

As a private investigator, I never get involved in the type of crisis situations Dr. Lord describes. However, I do crisis intervention in many instances in which my job becomes that of a counselor, or perhaps the conscience of a client.

Here is a real-life example:

Not long ago, my phone rang about 1 A.M.

"Guess where I am?" the male voice asked. I recognized him almost immediately as a client I had just done some work for—domestic surveillance. His wife was fooling around. I caught her red-handed.

When I finished the case, I gave him the information I discovered: the man's name, address, place of employment, and so forth. I am required by North Carolina statute's to turn over my information to my client. Rarely does it lead to complications, but in this instance, I became a negotiator.

"Where are you?" I asked.

He was sitting outside the lover's house, telling me his wife's car was in the driveway, and there were no lights on.

"Why are you there?" I asked.

"I don't know," he told me, "but I have to do something. I want her to know that I know."

He was angry. His manhood was on the line. He felt threatened.

Under North Carolina statutes, I am not required to call the police unless I feel there is an immediate danger.

So we talked.

" Have you been drinking?"

"No."

"Have you taken any drugs?"

"No."

"Do you have a weapon?"

"No."

"Do you want me to come see you?"

"No."

"Do you want me to call anybody for you?"

"No."

"Who is watching your children?"

"They are home alone."

That was the key question. Stalling for time allowed my client to calm down. And, it jolted him to realize he was, perhaps, jeopardizing his children by leaving them home alone.

"Look," I said. "Whatever you are thinking of doing, it can't help the situation. You could go to jail, depending on what action you take. You could lose your job. What would your friends and family think?'

I always ask a client thinking of doing something irrational, "What if your actions were printed on the front page of tomorrow's paper? How would you feel about others knowing what you have done?"

I asked my client to go home. Act as if nothing had happened. Forget this "drive-by," and we could meet tomorrow. "Call me when you get home," I said.

About 30 minutes later, he called.

"How are your children?" I asked.

"They're fine," he said. "Thanks for listening."

I did not hear from him the next day or the next.

Eventually, he filed for divorce. My surveillance information was given to her attorney, and the civil case was settled quietly.

There are other examples of my role as a negotiator. A defendant in jail wants to call some friends to go talk to potential witnesses against him.

I have to talk him out of that, warning him that it is an additional crime to tamper with a witness—obstruction of justice.

In these instances, I do notify the attorney I am working for about the threat.

Often, when a case arises, the initial meeting is with the attorney, the client, and me. There is an initial interview, and then the attorney leaves me with the client to get the details—names of witnesses and records that might be available—things to help in the investigation.

Often, the client will say he is planning to leave town; get his friends to lie; create documents; or go see potential prosecution witnesses.

At that point, we have a long talk. No negotiation. I point out the legal pitfalls of what he is planning and the dire consequences if he follows through. Again, in these instances, I notify the attorney I am working for.

Dr. Lord is right in her assessments.

Get the offender or client to calm down.

Talk about the problems, solutions, resources, and consequences of irrational behavior.

Mirror, empathize, but do not rebut. Validate, for example, "I know how you feel. I would feel the same way if that happened to me."

Polygraph Use

James R. Walker

James R. (Randy) Walker is a retired Agent from the FBI. Mr. Walker has more than 10 years of experience administering polygraph exams for the FBI and private clients.

Mr. Walker currently operates Walker Investigative Services specializing in private investigations and polygraphing. He is a member of the American Polygraph Association, and he is licensed in North and South Carolina.

Introduction

The case began as many of them often do. The matter under investigation was a sexual assault on a minor child on the Indian reservation in Cherokee, North Carolina. The mother returned home after leaving her children in the care of her live-in boyfriend. That evening during bath time, the youngest child, a girl no older than two, told her mom that her "peepee" area burned. The mother looked at the girl's vagina and saw it was inflamed and irritated.

The next day, the mother took her daughter to the doctor. After examining the girl, the doctor concluded that some type of penetration had been attempted on the child.

In all cases of this type, family court and law enforcement officials have to be called in to investigate child abuse. The live-in boyfriend denied doing anything to the child. There were no witnesses. The boyfriend agreed to a polygraph examination. I was called to conduct the examination.

During my pretest interview, the boyfriend confidently and calmly denied any form of sexual abuse on the child. He denied attempting to penetrate the child's vaginal area with his fingers or any other object.

My relevant questions during this exam were, "Did you put your fingers into that girl's vagina?" and "Did you put anything into that girl's vagina?" He answered each question confidently, no and no.

I read the polygraph chart. The boyfriend was lying. I told him the result.

"Can't be," he said, "I didn't do anything to that girl." His initial reaction was a somewhat faked appearance of surprise and more denials.

I continued questioning him. Gradually, I could see his posture sinking as he was sitting and a gradual acceptance of the inevitable. His continued attempts at denial were getting weaker and weaker. He needed time alone.

I left and spoke to the case agent. The man failed, I told the agent. He was nearing a breaking point. I needed a little more time alone with him.

I went back in and immediately resumed questioning him.

"She pooped," he said. "She made a mess. I was changing her diaper; she was crying; thrashing about and getting—all over me.

"I got angry trying to clean her," he said. "I finally thrust my finger into her vagina in a fit of rage."

The case agent came in. The offender repeated his confession. Charges were filed. Before the trial began, the boyfriend pled guilty.

Since the dawn of civilization, mankind has sought ways to distinguish truthfulness from lying, especially in those individuals suspected of criminal wrongdoing. Various techniques for the verification of truth and the detection of deception have been tried over the centuries, many ridiculous and cruel. Despite their primitiveness, each technique was based on the assumption that some form of physiological reaction occurred within a person when confronted with certain stimuli regarding a specific event under investigation, and this physiological reaction would, in turn, be manifested in certain recognizable external symptoms that were indicative of honesty or deception.

It has been reported that the Ancient Hindus required an accused person to chew a mouthful of rice and then spit it out on a leaf from a sacred tree. If the person could spit the rice, he or she was declared honest. If the rice stuck in the mouth, the person was lying. This test presumptively relies on the physiological response that makes a person's mouth dry when deceptive (Galianos, n.d.).

Early Bedouin tribesmen accused of wrongdoing in the Middle East supposedly had to touch their tongue to a red hot metal plate. If their tongue was not burned, they were honest. If the tongue was burnt, they were dishonest (Galianos, n.d.). This relies on the same physiological response of the person's mouth being dry when deceptive.

During the witch hunts in Salem, town elders had a "foolproof" way of determining if a woman accused of witchcraft was lying. They would tie her to a long, wooden pole and force her underwater. If she was a witch, she could save herself. If not, she drowned, proving she was not a witch (Galianos, n.d.).

In the 19th century, Italian criminologist Cesare Lombroso developed an early device for measuring and determining the pulse and blood pressure of a person undergoing interrogation, similar to the cardiosphygmograph component of the modern-day polygraph (Galianos, n.d.).

In 1921, William M. Marston, a student in experimental psychology at Harvard, invented a polygraph machine. Modifying Marston's polygraph machine, John Larson, a police officer in Berkeley, California, developed a technique for continuous recording of physiological responses. Then one of Larson's colleagues, Leonarde Keeler, added the gavanograph component to the polygraph and later established the Keeler Polygraph Institute of Chicago (Galianos, n.d.).

What Is a Polygraph?

The term *polygraph* literally means many writings. The name refers to the manner in which selected physiological activities are simultaneously recorded. Polygraph examiners may use conventional instruments, sometimes referred to as analog instruments, or computerized polygraph instruments. During the test, the examiner and the subject are alone in the questioning room. The subject is monitored by a polygraph machine and questioned by an examiner trained in forensic psychophysiology (American Society of Testing Measures, 2005).

The polygraph machine collects physiological data from at least three systems in the human body. Convoluted rubber tubes are placed over the examinee's chest and abdominal area to record respiratory activity. In other words, is there any detectable change in breathing?

Two small metal plates, attached to the fingers, record sweat gland activity. Is the subject beginning to perspire? A blood pressure cuff or similar device detects cardiovascular activity. Is the heart beating any faster when asked certain questions?

There are a number of different polygraph formats that have been published as a result of different polygraph research. The two most common formats are Control Question Technique and Guilty Knowledge Test (Elaad, 1999; Kleiner, 1999). The different formats may require more questions than others and a different sequencing of the questions. All formats require at least two and as many as five relevant questions.

The different formats also may have differences in how many comparison questions are asked. All require a certain number of irrelevant questions. Some allow an introductory question of the issues to be tested. The different formats also may have different scoring rules in evaluating the responses and rendering opinions.

A typical polygraph examination includes a period referred to as a pretest interview, the chart-collection phase, and a test-data analysis phase.

In the pretest, the polygraph examiner spends about an hour talking with the subject, completing required paperwork, and explaining the testing procedures with the examinee. The examiner uses this time to assess the emotional state of the subject and to develop the questions to be asked during the actual test. Before the test begins, the examiner goes over each question with the subject so he or she knows exactly what to expect. Contrary to popular belief, there are no "gotcha" questions during a polygraph exam. The subject knows all of the questions to be asked.

When they are ready to start, the examiner attaches the various components of the polygraph instrument to the examinee. The polygraph test itself usually consists of about 10 to 12 questions that require yes or no responses—no explanations and no elaborations. There are only two possible answers: yes or no.

During the chart collection phase, the examiner asks the predetermined questions, the components measure any changes in the subject's internal mechanisms, and these changes are then recorded on various polygraph charts—in other words, many writings.

Following this, the examiner analyzes the charts and renders an opinion as to the truthfulness of the person taking the test. There are only three possible conclusions:

- *No deception indicated.* That means no specific, consistent, and/or significant responses were observed at the relevant questions. The examinee passed.
- *Deception indicated.* This means that specific, consistent, and significant responses were observed at one or more of the relevant questions. The examinee failed.
- *No opinion.* This means that based on the data collected, the examiner could not make a determination. The examiner, when appropriate, will offer the examinee an opportunity to explain physiological responses in relation to one or more questions asked during the test.

I do not believe there are grades of deception. A person is either judged to be lying, not lying, or the results are inconclusive, and no opinion can be rendered.

Generally speaking, while there are exceptions, all persons being administered a polygraph test are told that polygraph exams are voluntary. FBI agents can be ordered to take a polygraph examination in noncriminal matters such as Internal Affairs (IA), Office of Professional Responsibility (OPR) investigations, or other administrative matters. Agents who refuse to take the polygraph examination face consequences—up to and including being fired.

Most agencies and individuals use some type of written form to advise the examinees and to have them acknowledge their understanding in writing. A written document is not always required; however, it is a good practice to protect the examiner from lawsuits by having the information in writing.

Purposes of Polygraph Examinations

Polygraph examinations have several purposes:

- Pre-employment screening
- Personnel security screening
- Criminal investigations
- Counterintelligence and counterterrorism investigations
- Misconduct and internal affairs investigations
- Source or informant verification–reliability

These uses of polygraph examinations are briefly described in the following text.

Pre-Employment Screening   Pre-employment polygraph examinations are nonspecific, full-scope examinations that are used to identify past behavior (e.g., use of illegal drugs, undisclosed criminal activity, omissions of information, falsification of employment application) that may indicate a lack of reliability in the potential employee. The Employee Polygraph Protection Act (EPPA) mandates that at the end of the exam, the applicant must be told the results and be given a chance to discuss the reading. EPPA exams are strictly regulated by the U.S. Department of Labor; so this is a federal law that covers private examinations. Not wanting to risk civil suits, many federal agencies just tell the applicant "you are no longer under consideration." There have been a number of lawsuits as a result of a failed polygraph, but I am not aware that any applicant has ever won his or her case in court.

The FBI precludes the examiner from an opinion on the exam immediately upon completion until it has been reviewed by a quality-control supervisor. Only at the end of that review is an official opinion rendered. The FBI does tell an applicant that he or she is not being hired as a result of a failed polygraph. An FBI applicant can appeal to the Special Agent in Charge (SAC) at the location of the original exam and request a second polygraph exam by a different examiner.

Most law enforcement agencies require applicants to pass a polygraph examination as a prerequisite to employment. Within the FBI, approximately 20% of the applicants tested are disqualified as a result of the polygraph. While studying at the American Polygraph Association (APA) Research Center at Michigan State University, Sgt. David W. Knight of the Flint Department of Public Safety (2000) conducted a study titled *Is the Polygraph Examination a Dying Investigative Tool in the Law Enforcement Community*. As part of that study, Sgt. Knight contacted police executives across the United States to determine the extent of, and conditions in which, polygraph testing was being used for pre-employment screening. The survey population included 699 of the largest police agencies in the United States, excluding federal agencies, and produced usable returns from 626 agencies. Sgt. Knight found that approximately 25% of the applicants tested by law enforcement were disqualified as a result of the information developed during polygraph testing, which is used to verify information provided in a job application and to develop information that cannot be obtained by other means.

Personnel Security Screening

Polygraph examinations are used in personnel-security programs to identify individuals who present serious threats to national security and to deter unwanted behaviors such as espionage.

Personnel-security polygraphs are nonspecific, counterintelligence-scope examinations. The examiners ask questions pertaining to the subject's involvement in espionage, sabotage, terrorism, unauthorized disclosure of classified information, and unauthorized foreign contacts. Personnel security screening examinations can be conducted as a condition of initial access to national security or sensitive information, as a part of periodic security reinvestigations, or on randomly selected individuals working within or with governmental agencies.

Criminal Investigations

Polygraph examinations used in criminal investigations are specific-issue examinations that are administered to subjects, witnesses, or informants to

(1) detect and identify criminal suspects, (2) verify information furnished by an informant or a witness to establish or corroborate credibility, and (3) obtain additional information leading to new evidence or identification of additional suspects, witnesses, or locations.

Every law enforcement agency has individuals who walk into their department on a daily basis wanting to speak to a representative and to provide some type of information. The law enforcement official has to evaluate the individual and the information given to determine how to deal with the information being provided. Information about a spaceship landing and little green men running around can be quickly dismissed. Information of more significance cannot be dismissed so easily and may need a quick analysis to determine the appropriate response.

Almost all federal law enforcement agencies have their own polygraph unit with experienced supervisory agents who oversee its polygraph program. One of the purposes of that unit is to have trained personnel conduct a quality control (QC) review of each and every polygraph conducted by that agency. The person conducting the QC review first reads the summary of the facts of the case as written and submitted by the original polygraph examiner. Next, the QC supervisor reviews the questions that the examiner asked to ensure that the appropriate relevant and comparison questions were included and properly formed. Without proper questions, the exam is invalid. Finally, the QC supervisor reviews the polygraph charts without reading the examiner's opinion so the supervisor can make an independent assessment and render an opinion based on his or her review of the charts. The supervisor would then look at the opinion written by the examiner to determine if their opinions agree. The opinion of the QC supervisor is the final opinion because he or she normally has more experience and training in polygraphs. Rarely are they in disagreement.

Those practitioners using the polygraph in the civilian community have various policies. Some do not have anyone review their polygraph charts. Others have a mandatory policy that all charts are reviewed by an independent examiner without knowing the opinion of the initial technician. In my private practice, I have every exam reviewed by another examiner.

At times, I will review the charts of another examiner, usually when those charts have been submitted by a defense attorney to a prosecutor asking him or her to change prosecutorial actions. In other words, the defense attorney is telling the prosecutor, "My client is innocent. Read the charts yourself. Drop the charges."

Not long after the World Trade Center disaster on 9/11, a person walked into the American Embassy in London asking to speak to the FBI attaché. This person reported knowledge of a plot to blow up a structure in London. If this information could be corroborated in any manner, a massive response would be put in motion by London authorities in an attempt to prevent this catastrophe. The FBI dispatched a special agent polygraph examiner on a plane to London to interview this person and to conduct a polygraph examination regarding the information he was providing. This person failed his polygraph examination, and during the posttest interrogation, recanted the information he was providing. This use of the polygraph saved the London authorities from a massive response and the wasted resources of their law enforcement manpower (Encyclopedia of Espionage, Intelligence, & Security, n.d.).

My most recent and significant case resulted from a telephone call from a defense attorney for whom I had previously worked. This attorney told me he had a client in jail awaiting trial on charges that he killed two men in retaliation for a past assault. If convicted, he could have been sentenced to death. I was told there was no physical evidence against this defendant. The only eyewitness testimony had been given almost 3 years after the shooting and from a person who was now in federal prison serving 15 years on a drug conviction.

"Test my client," the attorney said.

I spent almost 2 hours in my pretest interview with this defendant as we fully discussed his knowledge of the shootings, the people killed, and the informant. He denied being involved in the shootings and even being present when the shootings took place.

During the polygraph test, I asked the defendant, "Did you shoot either one of those men?" and "Were you present when those men were shot?" The results of the examination indicated no deception. Passing the polygraph examination indicated that he had not shot the men, and he was not present when the men were shot.

Do the results of the polygraph examination prove that he did not shoot the two men? No. Could he have passed the examination even though he was guilty? Yes. This case is pending. The defense attorney has given the results to the prosecutor requesting a review of the evidence.

Polygraph examinations may also be used periodically to verify information received from sources or informants. The practical investigator who is pressed by a heavy case load must rely heavily upon his sources of information because a good proportion of cases are solved by means of information supplied by informants.

While information often is voluntarily offered by those whose motives spring from good citizenship, others' motives may be fear, avoidance of punishment, jealousy, self-aggrandizement, financial gain, or revenge—all of which may encourage people to exaggerate or misrepresent the truth. Investigators should continually evaluate their informants and form an opinion of their reliability. The information received should be tested for consistency by checking against information obtained from other sources. A polygraph exam is a good method to test the veracity of the sources when there is no other way to corroborate information. It also lets informants know they will be checked, and their information corroborated.

Counterintelligence and Counterterrorism Investigations

During investigations into suspected security breaches (espionage) or foreign or domestic terrorist threats to national security, specific-issue polygraph examinations are administered to the subjects, witnesses, or informants associated with the incident or threat under investigation.

In early 1985, a disheveled woman, obviously an alcoholic, walked into the FBI's office in the Boston area wanting to speak to someone about her husband. She ranted that her husband was not paying child support as ordered. She blurted out that her husband was in the U.S. Navy and had been selling military intelligence information to the Russians.

There was some initial reluctance to give much credence to her comments due to her intoxicated state, except that she made references to some particular intelligence information that had to have come from a credible source. A skilled polygraph examiner familiar with intelligence matters tested the woman. She passed the polygraph, supplying intelligence information to the examiner that could be corroborated. What resulted was one of the largest and most significant spying and military intelligence breach cases investigated by the FBI. This lady's husband was John Walker, who had ultimately recruited a colleague and his son to perpetuate this crime. All of them were prosecuted and sent to prison (Encyclopedia of Espionage, Intelligence, & Security, n.d.).

The polygraph examiner, in addition to administering the polygraph examination, must be a skilled interviewer and interrogator. He or she elicits additional investigative information from the person polygraphed and often gains guilty admissions and confessions during the posttest phase of the examination that are useful to the investigator.

Misconduct and Internal Affairs Investigations

Specific-issue polygraph examinations may be given to employees who are subjects, witnesses, or complainants in investigations of personal misconduct in the performance of official duties. These examinations are used to substantiate or refute allegations, to verify information furnished by complainants or subjects, to establish or corroborate credibility, and to obtain additional information.

Polygraph Reliability

There is much debate as to the accuracy of polygraph tests. Most forensic psychophysiologists agree that the rate of detecting deceptive behavior is greater than the rate of detecting truthful behavior. This is because there are lots of situations where a defendant may have a guilty mind (and thus show up as deceptive on the polygraph results) but not, in fact, be guilty.

The APA claims that the accuracy rate for polygraph tests is between 85% and 95%, while others report accuracy rates as low as 70%. Why the difference? In making its assessment, the APA excluded those examinations where a "No Opinion/Inconclusive" result was found. Other research studies have counted those "No Opinion/Inconclusive" results as errors, resulting in the difference in the calculations of reliability.

So, the question arises, if these tests are so accurate, why are they not required in every situation where truth is at issue? Why spend billions of dollars on jury trials and independent prosecutors when we could just "wire up" the key witnesses and get at the truth with no fuss and no muss?

The U.S. Supreme Court recently decided a case that deals with this very subject. In *United States v. Scheffer* (1998), the Supreme Court upheld a military court evidence rule (Rule 707) that prohibits the use of lie detector results in military trials. In this case, Scheffer, a military investigator, took a routine urine test. The test came back positive for amphetamines. Scheffer then asked for and was given a polygraph test, which showed he had no knowledge of amphetamine use. At his trial on drug use charges based on the urine test, Scheffer tried to introduce evidence of his favorable lie detector results. The court refused to admit this evidence on the basis of military evidence Rule 707. Scheffer appealed on the ground that he should have been able to introduce the test results as part of his constitutional right to prepare his defense. The Supreme Court upheld Rule 707 for these reasons:

- There is too much controversy about the reliability of lie detector test results.
- Lie detector tests might undercut the role of the jury in assessing witness credibility.
- Lie detector tests create the possibility of vexing side issues about the reliability of the test.

That is why rules of admissibility during a trial differ from state to state. In most instances, the judge decides. In some instances, a polygraph may be admissible in court by stipulation and agreement of both the prosecutor and defense. In North Carolina, the Supreme Court has outlawed the admission of polygraph results regardless of any agreement between the prosecution and the defense.

In federal court, the Federal Rules of Evidence state when and if a polygraph may be admissible under the Daubert Rule, but trial use is still limited because the scientific community still has great reservations about polygraphs. The Daubert Rule is a particular case which addresses the introduction of any type of scientific evidence into a trial (*Daubert v. Merrell Dow Pharmaceuticals, Inc.*, 1993).

"Beating" the Polygraph

Polygraph examiners are always confronted by individuals who are concerned about someone's ability to "beat" the polygraph. The polygraph community defines these as *countermeasures*. These measures are defined as deliberate attempts by an examinee to use various means to beat the machine. They can be physical, mental, or chemical and vary from crude to sophisticated. The issue of countermeasures has become a matter of concern within the polygraph community because they can alter the results of examinations if not detected.

With the growth of the Internet, information on countermeasures and how to implement them is readily available to anyone. Websites such as Antipolygraph.com and Polygraph.com are devoted to convincing people that the polygraph does not work and that anyone taking a test, whether they intend to lie or not, should use countermeasures to help them pass. These sites also provide detailed information on how polygraph examinations are conducted and how to defeat the test.

Countermeasures fall into several categories:

- Physical. Using deliberate physical actions such as breathing, muscle contractions, and self-inflicted pain
- Mental. Using mental gymnastics such as disassociation, imagery, and biofeedback
- Chemical. Using drugs prior to the examination

In an effort to combat countermeasures, the federal government and other members of the polygraph community, as a byproduct of considerable research, have developed various techniques to identify and neutralize countermeasures. Physical countermeasures (the most frequent form attempted) leave a "finger-print" that is often easily identified by the experienced examiner.

Polygraph examiners are frequently asked about pathological liars and psychopaths and the belief that those individuals can beat the polygraph test at will. Psychopaths, now classified as individuals with antisocial personality disorder, do not have any remorse or guilt for their crimes or sympathy for their victims, and they do not accept responsibility for their own actions. The polygraph community conducted research studies that compare the polygraph testing of individuals clinically diagnosed as psychopaths with another group of individuals not classified as psychopaths.

The first well-designed study on the ability of some to beat the machine was conducted in 1978 in British Columbia, Canada (Hare, 1978). One of the principal scientists performing the study was Dr. Robert Hare, a world-renowned expert in the study and identification of psychopaths. The study found there was very little difference in accuracy during the testing, and the psychopaths were accurately detected at a slightly higher rate. During this study, no subject, psychopath or otherwise, was able to beat the test, and there were no *false negative* errors. False negative means those individuals who beat the test but were actually deceptive. Since that experiment was published, other studies also have shown that psychopaths and other poorly socialized people do not have an advantage when it comes to beating the test.

Why is this?

It may be that they have the same fear of detection and its consequences as anybody else. Also, it may be that the responses of our autonomic nervous system are so strong they cannot be overridden even by somebody with no conscience.

(I am often asked, "Doesn't the CIA train agents to beat the test?" Any information about training by the CIA, if it even is possible to train someone to beat the machine, is classified.)

Perhaps the most notorious case of a person beating the examiner and the machine is that of former CIA Agent Aldrich Ames, who for years spied for the Russians and was handsomely paid. Both Ames and his father were long time members of the CIA and very well respected.

Ames took two polygraph exams. In the first exam, Ames beat the examiner and QC supervisor. The examiner concluded Ames was not a spy, but he had problems with the question about money. It was money that should have caused anxiety for Ames as a result of all the money he was receiving from the Russians. After the fact, the FBI reviewed the first polygraph exam and concluded that Ames had failed. The second polygraph exam Ames took was a nondeceptive exam. Experts believe that since Ames passed the first exam, he had a higher level of confidence and felt comfortable taking the second exam (Encyclopedia of Espionage, Intelligence, & Security. (n.d.)).

People often ask about Robert Philip Hanssen, the former FBI agent who spied for the Russians for more than 20 years. How did he beat the machine? Believe it or not, Hanssen never took a polygraph during that 20-year span. After his capture, Hanssen received a life sentence, which he is currently serving in solitary confinement at the Supermax Federal Penitentiary in Florence, Colorado. (In 2002, *Master Spy: The Robert Hanssen Story*, starring William Hurt, was broadcast on television. In 2007, a commercial movie about Hanssen, *Breach*, starring Chris Cooper, was released nationally.)

It is not uncommon to find that an innocent person or one who is telling the truth fails a polygraph exam. Why does this happen? There could be a number of reasons: (1) a poor examiner who is overly aggressive in his pretest interview or does not ask the proper questions, (2) an examinee who has other issues going on in his or her mind that keeps him or her from focusing on the issue under examination, (3) extreme anxiety by the examinee that the examiner cannot control, (4) an examinee who may have been under the influence of alcohol or drugs at the time of the offense and is not sure what the truth is, or (5) a sociopath or someone that just does not care about lying and attaches no jeopardy to the issue of lying.

In the rare cases in which a guilty person passes a test, it is usually because of a poor examiner who does not ask the right questions. In these instances, it is the examiner who is being beat, not the test. The other instances in which guilty people could pass a test and be guilty are if, in their minds, they did not commit the crime. When answering the question about whether they committed the crime, in their minds they truthfully believe they did not commit the crime, so in their minds, they are being truthful.

Computer Voice Stress Analyzer

Various investigative techniques for detecting deception have appeared in the past 80 years. Some were developed by scientists and researchers, like reaction time tests, the polygraph, and brain-wave methods. Others were proffered by manufacturers without the help of researchers, such as the B & W lie detector and the various voice stress devices. The most recent method being heralded as "the new lie detector" is the computer voice stress analyzer (CVSA).

What separates the CVSA from previous voice stress methods is that the display is on a computer screen versus a printout. There are no validated algorithms, scoring systems, or sophisticated analytical methods. These shortcomings have not prevented the manufacturer from making remarkable claims regarding the efficacy of its product. Those of us in the detection-of-deception profession would like to believe their claims because using this new technology would allow us to better serve our clients and agencies in a shorter time.

Voice-analysis devices are designed to capture voice changes that may be related to stress in the act of deception. The voice-stress analysis is based on the theory that there is an inverse relation between stress and the frequency modulation in the human voice, and the voice-analysis devices are capable of detecting imperceptible changes in the frequency modulation.

Voice-analysis devices also have been tested to determine whether or not they could actually measure the level of stress. Although voice-analysis devices were able to detect stress under limited conditions, the performance was not consistent, and overall results did not provide convincing evidence. The laboratory studies consistently showed that the standard polygraph method performed better than the voice stress analysis, and the voice stress analysis did not perform better than the chance level.

The Virginia Board for Professional & Occupational Regulations (2003) conducted a study to make a determination about allowing the use of voice stress analyzers. The board wrote that a review of the current literature and summarization of the four public hearing sessions and written comments uncovered a continuing polarized debate between the polygraph and voice stress communities. The conflict arose from the lengthy history and regulation of the polygraph compared to the mostly unregulated new technology of voice analyzer equipment.

There have been several scientific studies conducted on the polygraph over the years, and while no study has indicated the polygraph to be 100% accurate, it has still been deemed a reliable instrument to detect deception when used correctly. On the other hand, there has been no independent scientific evidence to indicate that the computer voice analyzer is a valid instrument to detect deception. The only evidence that has been presented and reviewed, to date, consists of testimonials and other anecdotal evidence. As a result, the Virginia Board recommended that computer voice analyzer equipment should not be approved in Virginia at that time.

Conclusion

The use of polygraph always brings forth a great deal of debate as to the efficacy of its utilization. Inside the FBI, we always considered the polygraph as a great tool in making decisions in how to proceed in investigative matters.

In the case of the person providing information, which is going to necessitate a significant and almost immediate response, the polygraph will provide confidence to law enforcement administrators that they are responding appropriately when the person passes the polygraph, and they can "stand down" from responding if the person fails the polygraph.

During the course of criminal investigations, persons may come under suspicion of criminal wrongdoing and may be asked to agree to a polygraph examination. If he or she passes the polygraph, that suspicion is removed. If the person fails the polygraph test, the investigation continues. During plea negotiations between the prosecutors and defense attorneys, the polygraph is often suggested as a means to clarify what a subject is accused of and what he or she has actually done so that an appropriate plea is determined.

Research continues within the academic and scientific community looking for a better means for testing and determining the truth. While some methods, such as a brain MRI, have shown promise, nothing has yet been shown to have the reliability or effectiveness to replace the current polygraph test.

ALLEN'S WORLD

Rusty had been convicted of "shooting into an occupied dwelling at night," a felony in North Carolina with a mandatory life sentence. He was in Central Prison in Raleigh without hope of ever being a free man. As a reporter who covered Rusty's trial for *The Charlotte Observer,* I was convinced Rusty was innocent.

Rusty was a member of the Outlaw motorcycle gang. Not the most celebrated member of our community. But I covered the trial, did not think he had a fair hearing, and wanted to see that justice was done. Rusty was in California with his wife the night of the shooting, and I could prove it. I was about to win a Pulitzer. Before we published my evidence, my editor, Walker Lundy, insisted that Rusty take a polygraph exam.

District Attorney Peter Gilchrist III made a deal with *The Observer* editors. If Rusty passed the exam, Gilchrist said he would move for a new trial and drop the charges. If Rusty failed, too bad. He would stay in prison the rest of his life.

Rusty failed.

I implored Gilchrist to test Rusty's wife, who was with him in California. Gilchrist did.

She failed.

I visited Rusty in Central Prison. I wanted to find out how he could have failed the exam. The warden, Sam Garrison (now deceased), took me to the hospital section of the prison. I looked through a glass viewing window into Rusty's cell. Rusty was naked. He had stripped his bed of his sheets and the mattress and was bouncing up and down, on the springs, shadow boxing. I did not go in to see him.

I am convinced to this day that an innocent man is in prison for a crime he did not commit. I believe that Rusty has what I call "guilty knowledge."

I believe he knew the assault was going to take place and did nothing to stop it. I believe Rusty was guilty of other, more serious crimes and, perhaps nervous that something else might surface, failed the exam.

As a private investigator, I rarely get involved in polygraph exams. That is a determination made by the attorney I work for. I do, however, circumstantially deal with polygraph testing.

For example, John called recently. He was about to apply for a job. A polygraph exam was mandatory prior to employment, and he wanted to know how they worked.

"What's the problem?" I asked.

Typical questions on a pre-employment exam center on shoplifting, the prior use of drugs, or any criminal charges regardless of the outcome.

John told me that about 15 years ago he was charged with marijuana possession. When asked about drug charges during the exam, John wanted to know, did he have to make that disclosure. I am not an attorney. I never give legal advice. So John put me in a quandary. What do I tell him?

I told John that if that charge was still on his record, I would find out. If I could find it, I told him, the examiner and the company he wanted to work for could find it. The decision to disclose it would be up to him. All I could do was tell John if it was still on his record. So John told me when, where, and the disposition of the charges. I searched all available computer databases and called the county of record. All paperwork on John's case had been destroyed. In other words, there was no record anywhere on this planet of the old charges against John. I passed that information to John. I never followed up on how he handled the job exam, or if he got the job.

District Attorney Gilchrist, who's been on the job since 1975 (Gilchrist has announced that he is not running for re-election), believes in the reliability of polygraphs depending on the circumstances. Who administered the test? What preparation was there for the test? What questions were asked? How were they phrased?

Gilchrist said in an interview for this chapter, he will not accept the results of a test brought in by a defense attorney.

"No matter who ran the test," Gilchrist said, "I would be very suspect. I want the polygraph operator to talk to me and/or the investigator who worked the case, rather than have a man working for the defense phrase his own questions. I would give no credence to that."

Gilchrist, will on occasion, ask that one of the police experts administer a test to help determine which way to go during an investigation. A cooperative witness or defendant can gain points by voluntarily agreeing to be tested, while people who refuse consent might be viewed with skepticism, which does not always mean they are guilty.

Take the case of the judge's daughter accused of passing bad checks. Gilchrist remembers that the woman was picked out of a lineup by the clerk who took the check. Gilchrist offered the judge's daughter the choice:

take the test. If you pass, we will drop the charges. If you fail, we will use that result in court. The woman refused to take a police-administered polygraph. She hired her own technician, passed that test, and asked Gilchrist to drop the charges. He refused. The case went to trial. The woman was acquitted.

Then, there was the case of a man charged with rape. The victim told police she was walking down the sidewalk, and a man jumped out from behind a bush, dragged her back behind the bush, and raped her. Police arrested a suspect. The victim positively identified him. "That's the man," she said. The rape took place before DNA testing. So the case boiled down to her eyewitness identification. At trial, the defendant brought in two alibi witnesses who said he was with them when the rape occurred.

"It was obvious they were lying," Gilchrist said. The man was convicted and sent off to prison. Shortly after, the defense attorney found a picture of the man just after he had come out of military service. His appearance was drastically different from the physical description given by the victim.

"My client is innocent," the defense attorney argued to Gilchrist.

Gilchrist said that argument, and the picture, created doubt.

"I'll make you a deal," Gilchrist told the defense attorney. "Have your client take a polygraph administered by one of my technicians. I'll go to the judge if your man passes and ask him to set aside the verdict."

The defendant took the test. He passed. The man was released. The rape case has never been solved.

I do not think the polygraph ever will be routinely admitted in court. It is a tool, a means to an end. I asked Gilchrist if science and the use of polygraphs as an investigative tool serve as a means of justice.

"It is justice being done," Gilchrist said, "as well as I can do it."

13
CHAPTER

Stance-Shift Analysis

Boyd Davis, PhD, and Peyton Mason, PhD

Dr. Davis is professor of linguistics in the English department at the University of North Carolina-Charlotte. Dr. Mason is the chief executive officer (CEO) of Linguistic Insights, Inc. in Charlotte, North Carolina.

Introduction

Words are slippery. Just ask any lawyer, investigator, prosecutor, linguist, or communications scholar. Humans begin manipulating words and meanings as children. That manipulation continues throughout adulthood. Manipulation never ends regardless of the time, place, or circumstances. Here, Nikki is suddenly on the spot:

"Nikki, what happened to that glass?"

"It broke, Dad. It just got broken."

Did Nikki break the glass? Was it her little brother? Or, was it that darned cat that actually set in motion the train of events leading to somebody dropping, pushing, or knocking over the glass? We cannot tell. The cat is not confessing anytime soon. All we know for certain is that someone or something broke that glass.

Yet, we can begin building a hypothesis before asking questions.

Once we have answers, we can begin to draw inferences. Once we draw inferences, we can use the stance-shift analysis technique to find deception or evasiveness and prepare follow-up and/or probing questions.

What is stance-shift analysis? In very simple terms, it is a technique that uses computer software to identify shifts in speech—after careful analysis of a written transcript or statement—to guide the interviewer to places that he or she needs to go back and probe. Stance-shift analysis is not used during the actual oral interview.

The example of the broken glass may seem trivial. However, the interpretation and inference-building process involved in analyzing an interviewee's statement is complex, so much so that investigators often turn to external systems of analysis to find a blueprint for further probing.

Interviewers, attorneys, prosecutors, and investigators who are all so confident in what they are doing, think they are infallible. They all are sure they will spot the deceptive phrases in an oral interview, or they will spot the evasive responses. The best interviewers on this planet miss stuff, and that is when stance-shift analysis enters the picture.

Stance-shift analysis pinpoints the areas that need to be probed. Here is a more serious, real-life example of how stance-shift analysis paid off. The example is intentionally vague to protect our client's confidentiality.

Nine depositions had been taken by an attorney. Each respondent was asked the same questions. In reviewing the transcripts, the attorney accepted the fact that the respondents were accurate and factual, but something bothered the attorney. So, he brought the transcripts of the nine depositions to us (authors Davis and Mason). Using stance-shift analysis, we saw a pattern. On one question, each respondent showed signs of possible deception or, at the least, evasiveness. Our report to the attorney highlighted the one question that needed probing. He re-interviewed the nine respondents, focused on that one question, and turned the case around.

In this chapter, we will guide the students through the various styles and different approaches to statement analysis as used by various investigators, psychologists, linguists, and others seeking information from a respondent. This book has already looked at specialty areas to spot deception, such as investigative discourse in Chapter 6 and the polygraph and voice stress analysis in Chapter 12.

Statement Analysis

Susan Adams (1996) has taught statement analysis for years at the FBI Academy in Quantico, Virginia. Adams directs interviewers to focus on nouns, verbs, and the balance between the various parts of a statement because "linguistic analysis techniques can assist law enforcement in determining veracity of statements and help detect sensitive areas that may need further investigation" (Adams & Jarvis, 2006, p.1).

Statement-Validity Analysis This technique began as a tool to assess testimony by child witnesses in sexual-offense trials and included a semistructured interview format (Koehnken, 2004).

Criteria-Based Content Analysis This technique analyzes transcripts using 19 criteria and evaluates its scores through a validity checklist (Vrij, 2005; see also Shuy, 1998).

While statement validity analysis and criterion-based analysis are readily acceptable as techniques for good listening for interviewers, there is considerable debate in the field of forensic linguistics as to their reliability in identifying areas of deception and veracity in written statements of interrogations, depositions, and testimony. Also, there is professional debate as to their usefulness for investigations or admissibility in court as scientific evidence (Leo, 2008).

Automated Techniques Hancock, Curry, and Gorham (2008) summarize some automated techniques keyed to the investigators' assumptions about deception and deceptive language, and typically suggest that four categories of linguistic cues can be identified as involving deceptive messages:

1. Word quantity. A belief that a deceptive or evasive person will use more words in response to questions than a candid person uses.
2. Pronoun usage. Some respondents use pronouns to "catch the interviewer's attention." When a pronoun seems out of place, or is "curiously missing," the interviewer needs to probe. Pronoun usage is a way some respondents have of distancing themselves from an event. As mentioned in Chapter 6, use of pronouns can be used to depersonalize a reported relationship, such as always using *she* for wife.
3. Emotional words. Words such as *love, hate,* and *despise*; words used by the respondent intending to deceive.
4. Markers of cognitive complexity. Words such as *because* and *since* might indicate what the speaker suggests as causative factors; those areas will need to be probed.

Gender Genie is a software-based technique that identifies the gender and age of unknown or anonymous writers of documents (Koppel, Schler, & Argamon, 2009). This technique is most useful when the author of a document, such as a ransom note, is unknown.

Linguistic Inquiry and Word Count is a software program developed by psychologist James Pennebaker to look at "empirically derived statistical profiles of deceptive and truthful communications." Pennebaker is looking for words that indicate "affective language." He believes affective language might indicate lying words or emotional words. The frequency of affective words in a transcript characterizes the tone of a document that might contradict the document's actual content, which lets the investigator compare and match styles (Hancock et al., 2008).

Fuller (2008) offers empirical testing of constructs identified by researchers associated with Arizona's Center for Management, directed in 2009 by Dr. Judie Burgoon, herself a noted deception analyst, whose affiliated researchers are working with audio, video, and online text analysis.

Stance-shift analysis is a more sophisticated, noncontroversial system of detecting potential deception or evasiveness in written documents, which includes transcripts of oral interviews, interrogations, conversations, depositions, or testimony. It is empirical in that it is based on a speaker or writer's personal usage.

The way a person says something is part of what that person means. We call that a *usage pattern*. People have their own patterns of word usage for the way they talk, although they seldom think about the fact that what they are doing while they are talking or how they say something often adds another dimension to the message they are intending to send.

Those usage patterns signal our attitudes about one or more of these features: what we are saying, the situation we are in, and the person with whom we are talking. For example, if we are not particularly confident about what we did or what we intend to do, we will signal that by using "might," for example, "I might have gone to the mall."

A sentence later, the interviewee commits to an action: "I know I stopped at the post office." We have gone from a stance of uncertainty to one of certainty. The interviewee has shifted his or her style of talking.

In stance-shift analysis, that shift in a person's style of talking has meaning. In talking to one another, people constantly position and reposition themselves by the way they address each other, or what they choose to reveal, or by the speech acts such as questions or interruptions. In one expression, speakers may present themselves as fearless adventurers. In the very next expression, they may shift to words and pronunciations that make them appear as fearful children.

Stance-shift analysis is our term for the ways people use words to signal where they obtained their information and how reliable they think it is (evidential); their appraisal or judgment about topics, situations, viewpoints, and values (evaluation); their intention to do or to not do something; whose responsibility the action will be (agency); and details about emotion (affect).

When we are listening to people, we focus on what they are saying, and only occasionally do we pay attention to how they are saying it. For instance, the pediatrician talks one way to the child and another way to the parents. A shift in stance is a shift in style. Tracking shifts in style gives stance-analysis professionals a window on how the people use language to act toward each other.

Stance-Shift Analysis

Instead of talking about deceptive language, we should be talking about the speaker's change in the use of language used to deceive or mislead. The language does not change—it is the speaker's usage that changes. That is one reason that a number of researchers have worked on automating the analysis of interviews and interrogation, because they can use computational analysis to measure and analyze words as they are used.

Stance-shift analysis does not specifically identify deception, but instead focuses on what Kingston and Stalker (2006) would classify as forensic stylistics. When a person shifts stance, something is changing. During an interview or while giving a statement, the speaker might shift from an appeal to authority to a timid demurral or to a sudden backpedaling. The speaker has shifted stance.

Tracking the stance shifts lets the analyst see how the speaker is using language for different purposes, including repositioning the participants or the frame of reference of an interaction, which affects how the acceptance or projection of responsibility is presented to the interviewer.

Stance-shift analysis uses computer-supported coding for stance shifts designed to decrease bias in coding. As we said above, this is a complex computer program. Instead of going into the minutiae of how we developed the system and how it works, what follows is a basic explanation of how the system operates (see Lord, Davis, & Mason, 2008).

Our system incorporates the identification of at least 80 different language categories. Of those, some 56 are not used in our coding. That is because these 56 most likely contain fact-bearing words such as *bullet, car,* and *television.* While these words are important as informational words, they do not help us characterize a shift in a respondent's stance. So, that leaves 24 different language features that we have found to be significant. These contain words that researchers call *stance words.* Anyone who knows how to diagram a sentence will recognize some of the 24 language categories, which include different types of adverbs, different types of verbs, different types of adjectives, different types of pronouns, and other parts of speech (**Figure 13.1**). The more important words are given more weight than those that are not as useful for identifying the four dimensions of stance. Each of these 24 different language features is assigned a weight that indicates its importance in one of four dimensions:

1. Opinions: Phrases such as "I think" or "I believe"
2. Rationale: Close to opinion words or phrases, but more detailed (because . . .)
3. Feelings: Detailed mood statements
4. Agency–personalization: Distancing or assuming responsibility for an action

Our analysis isolates where the interviewee is revealing attitude that he or she might not have wanted to show or not have noticed. Each scale will indicate the degree of stance being produced as the respondent speaks.

The following illustrates the analytic steps on a sample of an oral interview transcript:

Q. What did you do after that?
A. I think I went to the post office. But I may have gone to the grocery store first, and then to the post office.

Figure 13.1 Stance Identifying Language Features

- Adverbs:
  additive (*also, too*)
  linking (*anyway, however*)
  conditional (*if, unless*)
  time (*afterward, soon*)
  degree (*exactly*)
  stance (*actually*)
  causative adverbs (*since*)
- Negatives (*not, n't*)
- Adjectives: (*bitter, cheap, rich*)
- Pronouns: First, second and third person; indefinite (*anybody*); impersonal (*it*)
- Verbs: public (observable: *walk*), private (*believe, feel*) and (per)suasive (*urge*).
- Modal verbs: possibility (*can*), necessity (*should*)

- Adverbs–adverbial phrases used as
  amplifiers (*completely*)
  emphatics (*really!*)
  downtoners (*barely, only*)
  hedges (*almost*)
  quantifiers (*some, all*)

- Discourse particles (*well*)
- Conjuncts (*else*)

Adapted from: Mason, P., B. Davis and D. Bosley (2005). Stance analysis: When people talk online. *In Innovations in e-marketing,* II, (ed.) Sandeep Krishnamurthy. (pp. 261–82). Hershey, PA: ICI.

Each question is identified by a Q, and each answer is identified by an A. We first separate the questions from the answers. We can analyze both questions and answers, though we focus initially on the answers.

Next we take the responses, and starting at the beginning, we divide the answers into segments, each containing 100 words. For example, if the responses contain 1,000 words, we will have 10 blocks of 100 words each, and each word is coded for a language feature. The language features that represent stance are each assigned a score.

All 10 blocks of 100 words each are fed into the computer to score the segments for the four stance dimensions. The last step is a chart that shows the scoring for each segment of words generated by the stance-shift analysis software (**Figure 13.2**).

The larger the score, the more stance is being expressed by the respondent. These are the areas in the text that identify where the questioner might need to go back and probe further.

Going back to the example at the beginning of Nikki and the broken glass, when asked who broke the glass, Nikki might respond:

"The glass broke." No agency.

"I broke the glass." Agency. Nikki is taking responsibility.

"Vivian broke the glass." Distancing and allocating responsibility to another.

Let us take a look at how stance-shift analysis works in the real world.

Mrs. Willow (not her real name) gave several depositions in a malpractice lawsuit she filed.

The lawsuit eventually settled out of court just prior to Mrs. Willow's death. We analyzed the depositions after the case was settled.

In most of the sections in which Mrs. Willow shifted stance, she was reporting her speech and the speech of others, using the verbs *know, say, speak,* and *tell*. Mrs. Willow made a clear distinction in terms of personal agency and action around those verbs. When she reported that a medical person had *said* something, she apparently assumed the medical person had delivered information or made a comment requiring no action on her part.

Figure 13.2 Stance-Shift Analysis Chart

Word Segment	Opinion	Rationale	Feelings	Agency
1	0.62	0.58	−1.00	−0.99
2	1.84	0.56	0.10	−0.86
3	2.36	−0.17	0.03	0.05
4	0.78	−0.63	0.48	−0.93
5	0.62	−2.00	2.92	−0.28
6	0.72	−0.74	−0.53	−0.88
7	−0.30	−0.87	1.51	−1.03
8	0.56	−0.07	−0.11	0.47
9	0.18	0.87	−1.06	−0.82
10	−0.72	0.42	−0.79	−1.32

In other words, Mrs. Willow did not think she had to take any action—that there was nothing she was required to do. In the depositions, Mrs. Willow does not describe any actions that she took in response to what anybody said. However, whenever she reported that a medical person *told* her to do something, the rest of the section and the next sections show her taking action.

Mrs. Willow interpreted personal responsibility when she was told to do something. This distinction between saying and telling may have contributed to her death. One point of contention was whether or not Mrs. Willow understood that remarks about colonoscopies went beyond the delivery of information and were actually intended to get her to do or ask something (we have boldfaced the say–tell verbs directly from the deposition):

> "In my colon? No one has **told** me that I do, and no one has **told** me that I don't since the surgery, when I was **told** I did not have cancer of the colon anymore. . . . And my CAT scans don't reflect cancer in the colon. I've been **told** that cancer could be undetectable, depending on size and depending on where it is . . . And I explained to her what my symptoms were and gave her some family history. And she immediately **said**, "Has your doctor ordered a colonoscopy?" And I started crying and **told** her that that's the same thing Dr.___ had asked me, and why is this such common knowledge and my doctor hadn't ordered it?"

Had Mrs. Willow understood that what was said to her required her to take some positive action, the outcome of the case, and her life, could have been different.

A second example is taken from a study of sexual crimes (Lord, Davis, & Mason, 2008). Mr. Hyde (not his real name) was interviewed and confessed to rape and murder. The stance shifts in Hyde's statement show how he pushes personal responsibility aside to justify the reasonableness of employing aggression. He substituted *you* or *we* for *I* to distance himself from actions or words or to project behavior onto others ("We had sex", see **Figure 13.3**), and he used *they* for unnamed audiences.

Hyde also used specific categories of markers such as conjunctions (but) to move between tentativeness and certainty. Typically, the direction of such movement may signal that rapists are reconsidering their personal responsibility for actions, and here, it contributed to when and how Hyde is presenting the murderer as different from the person he now considers himself to be ("Where he claims agency" in Figure 13.3.).

Hyde's full statement illustrates O'Connor's continuum of agency (2000) that runs from deflecting agency to problematizing agency and eventually to claiming agency (p. 44).

Again, agency identifies responsibility. In the sections of his transcribed interview that showed high weighted scores for agency, Hyde explained that his murderer-persona committed the crime (deflecting personal responsibility or distancing) and then claims, when his victim chose to run away from him, that action caused her death (deflecting personal responsibility; see "Where he claims agency" in Figure 13.3).

Figure 13.3 Summary of Content in Significant Sections

Where he is extra careful	His knife has an extra sharp edge; Victim's earring: was it found? (he introduces question); "**We** had sex" [note pronoun]; Bragging rights: about sex; in prison. Bad company in youth. Others w/pickup trucks confused with him.
Where he "misremembers"	What he wore; what she looked like. Details of sex act. Admits stabbing but claims no memory of what happened afterwards until hiding victim [dissociative].
Where he elaborates	Power, mastery, control of victim; references to knife. Dedicated— he "had to do" the murder. How he "changed" later. Subtext: "2 wrongs"—he asks himself which is more wrong (murder or rape).
Where he claims agency	Confusing sections: when he did a murder, he created a murderer [who was then the responsible party]. Chased victim down: however, her escape effort is what brought about her murder [dissociative].

Adapted from: Lord, V., Davis, B., and Mason, P. (2008). Stance shifts in rapist discourse: characteristics and taxonomies. *Psychology, Crime & Law* 14, 357–79.

In the following section from Hyde's transcript, the *they* seem to be the victim's loved ones who cannot forgive him, although he wishes their blame would be focused on the event, not the person (problematizing agency). His use of a cause-and-effect is centered on the murderer, who no longer exists (claiming agency):

> ". . . and by that stupid act I've created somebody that could do almost the same thing I done, destroy a life, because if I could . . . (INAUDIBLE) . . . and **they** say hey I can't forgive it. If I was in **their** place I'd probably feel the same way, and that right there . . . that can (INAUDIBLE) . . . that's something I don't have to worry about. But think about what **they** do . . . a bunch has been taken from **them**. Can't . . . can't nothing undo it! Can't nothing change it. I just wished **their** focus would be what has happened and not at me so that **they** can say Lord . . . help me, and **they** could learn to put **their** focus on what happened. Nobody took it. 'Cause that man don't exist no more."

A third example showing how people signal affect with details is from an insurance deposition. The interviewer asked a question, got very little information, and then asked an open-ended question, letting the interviewee associate as desired.

In this instance, errors had been made in reinstating a person on insurance policies for property that suddenly burned. The interviewer is probing the extent to which the insurance agent's errors might have been careless, as opposed to deliberate. The insurance agent is concerned about presenting himself as incisive. As he talks, he begins to suggest an underlying concern:

Q: Do you recall what he said about that?
A: No, I don't, I don't.
Q: What else, if anything, do you remember about that conversation with G___ on the 28th of December, 2___?
A: I just remember thinking the whole thing from the very beginning was—you know, was strange, I mean, and very bizarre, because like I say, I had knowledge that it was going, you know, in a hostile manner. From talking—you know, you know, with both of them, I recall them making all these other different changes and things, you know, and wondering and really kind of trying to disseminate from talking with him if—you know, where he was coming from, you know—you know, what his—you know, if he was guilty of anything, you know, in the process. I mean, I was—and to this day, I mean, there again, there was a separate rental house that burned and within a day or two of when this one burned with no insurance on it.

In the above example, one cannot overlook the frequent repetition of "you know" and "I mean": these phrases surfaced only in significant sections where the speaker was given an open-ended question and began stumbling over his explanation and offering one detail after another. In a number of analytic techniques, this repetition would be taken as a sign of deception. Stance-shift analysis does not claim that this type of repetition is necessarily a sign of deception, because in a high-stakes interview, respondents often become nervous. Instead, stance-shift analysis identifies where the speaker shifts throughout the interaction, and the interviewer can use those areas for follow-up questions and further probing. In this interview, for instance, the speaker used "you know" and "I mean" phrases in responding to anything that asked for more than a yes or no answer.

Conclusion

Like other approaches that automate the analysis of text, stance-shift analysis seeks an empirical way to identify salient sections of an interview that will help the interviewer analyze the speaker's statements. People have multiple reasons for how they say what they say, and they may not be aware of all of them. Listening, asking, reviewing, and reflecting—these actions help the interviewer. As an automated technique, stance-shift analysis saves time and reduces observer bias—the chance that the interviewer is hearing on only one channel when the speaker is sending a message on more than one channel.

PART

IV

Integration

Putting It All Together

OBJECTIVES   Upon completion of this chapter, students should be able to

1. Identify a criminal justice professional to interview
2. Develop objectives for the interview
3. Develop questions that address the objectives of the interview
4. Predict and develop probes for the interview
5. Conduct a 15-minute interview with an identified criminal justice professional
6. Transcribe the interview
7. Analyze the interview using a set of structured questions

Introduction

Becoming an effective interviewer takes conducting many interviews and discovering what works and what does not work. Adopting and using the basic skills outlined in this text, each individual must discover what formula works for him or her.

As with most skills, interviews can be conducted hundreds of time and still result in mediocre outcomes. As much as possible, each interview must be prepared, and upon completion, each interview should be analyzed, asking the key question, "What could I have done better?"

This chapter will not add additional information. Instead, it will help the beginning interviewer plan a practice interview and provide a structure for analyzing the practice and future interviews.

Structuring the Practice Interview

The practice interview is an opportunity to rehearse interviewing in a relatively safe environment and to learn more about a profession that the student considers a "dream" job.

This practice interview is an opportunity that should not be squandered on students selecting the easy way out of an assignment. Everybody knows a neighborhood police officer or has an uncle who works for probation. That familiar professional will easily answer questions and is invested in helping the student interviewer. The student interviewer will be practicing, but will receive limited benefit. In other words, selecting an "easy" target does not help.

Identifying the Professional and Scheduling the Interview

The student should find a professional he or she does not know. This person should be employed in a job that the student believes is the ideal job.

If the student has always wanted to be a secret service agent, he or she should find one to interview. The stress generated by this person will be palpable, making it an interview that brings into play all of the techniques developed in this text. The students also will have the opportunity to find out more about the career that interests them.

The students should select a criminal justice professional who is stationed in their city so they can be interviewed in person and audiotaped. More than likely the academic institution has faculty who have research relationships with the professionals and can help identify specific individuals within the profession who will be willing to provide 15 to 20 minutes for the interview.

As with real interviews, students need to prepare suitable introductions for a successful scheduling of the interview. The introduction needs to include brief information about the student, the objectives of the interview, and the importance of interviewing the identified individual. Thus, the student has an opportunity to develop skills of motivating the interviewee.

Students need to explain to the professional that the interview will last only 15 minutes, and the interview must be recorded on tape. These two facts are important. The professional's time is valuable, and the fact that the interview will take only limited time will increase the probability of acceptance. The student must be able to transcribe the interview so taping is required.

The professional may not be comfortable with taping, or the agency may have a policy against it. The students should always ask before interviewing if taping would be a problem. To just show up assuming that the professional will be comfortable being taped is a guaranteed avenue to create ill will and a last-minute refusal.

Before the scheduled interview, the student should have a tape recorder, know how to operate it, and run a brief test to ensure that it works. A new tape should be used, not an old tape that must be taped over. It is possible that the underlying material might bleed through.

The students should be sure to schedule a time and location convenient for the professional. As noted in Chapter 4, they should be sure that the mutually agreed upon location is free of distractions and private. The interview should be scheduled during the term after the student has read and received instructions on interviewing, but with sufficient time to allow transcribing and analyzing the interview.

It should go without saying, but the students should be sure to thank the professionals in advance for their time.

Developing an Objective and Corresponding Questions

Students should develop the objectives for their interviews before contacting the professional so the students can articulate what they would like to discuss.

Some professionals want the questions before the interview. Unless requested, it is better if the professionals do not receive the questions. Having knowledge of the general objectives of the interview makes the professional comfortable with participating in the interview. However, too much preparation may result in "canned" responses.

The students should review each others' objectives and questions and then obtain additional feedback from the instructor. Exercises at the end of Chapters 3 and 4 will help provide additional direction.

As would be desired in actual interviews, the students should research the professional's agency and possibly even what has been written about the professional. An Internet search may reveal news articles about cases or clients with which the professionals have been involved. Prior information helps students develop rapport and also may include some areas for questioning or areas to avoid.

Although the definition of probes includes improvisation, formulating some probes that might be needed helps the students prepare and revise their questions. Reading each question, the students will think about the professionals' responses. If the students have researched the profession and the professional, they will have some idea of the answers the professional will provide. What might be responses that will need further clarification or elaboration? The students can brainstorm probes but then also consider whether or not their questions could be written more clearly.

Conducting the Interview

Students need to minimize their anxiety about the interviews. Practicing their introductions with other students or in class will be helpful. The introduction should include an appreciation for the professional's time and a rapport-building statement.

> "Agent A, I'm Allison Walker. Thank you so much for taking time out of your busy day to meet with me. I have always been fascinated by the responsibilities of Postal Inspectors. It's a career that I would like to pursue. The opportunity to ask some questions of somebody with as many years experience as you is a dream come true."
>
> "As I mentioned on the phone, I'll try to limit the interview to 15 minutes, and I'll be tape recording it. I believe you said that you had a quiet room that we could use?"

The students should be very clear on the interviewing location. After establishing by phone where and when the interviews will take place, students should write a letter or email to the professional confirming the agreed upon location, date, and time. Before the date, the students should be sure they are familiar with the directions to the location and leave plenty of time to arrive early. Dress is important. Remember that the interview is with a professional. The students should dress as if they are going to a job interview.

Dress is a sign of respect and consideration. Most professionals recognize the care the students have taken and will be more willing to invest time and effort into the interview if it appears the student is taking the assignment seriously. A slovenly appearance does not convey an attitude of seriousness.

While conducting the interview, students should remember what they have learned. It soon will be clear to them just how difficult it is to listen actively, paraphrase, evaluate for complete and relevant responses, and observe nonverbal cues.

When entering the interviewing room, students need to find a location to place the tape recorder. Ideally the tape recorder will be portable to allow flexibility. Make sure it can be placed where it will pick up both the professional's and the student's voices. Testing it is a good idea, and the professional will understand.

An additional challenge for students is to try to not talk over the respondent. Unfortunately, students cannot stop the respondent from doing the same. Transcribing is difficult enough without attempting to figure out what was said and who said it.

Students need to figure out before the interview how they will keep track of time. Time-keeping is an artificial step that should not be a component of future interviews. It is possible that the professional may even help the student keep track of time so the student does not end up with 30 minutes of interview to transcribe. Although not recommended, the students can end the interview after 15 minutes with "Fifteen minutes is up, but if you have a few more moments, I would like to hear more. I want to record how much I appreciate your time, and then I will turn off the tape."

No matter what, students should be sure to end the interview with appreciation for the professional's time.

Transcribing the Interview

A transcription includes every word that is said during the interview. These words include filler sounds such as "uh" and "you know." Silences of more than 5 seconds should be indicated. It also is useful to include such sounds as laughter.

On the transcript, students should use *I* to indicate each time they begin a question or comment and an *R* each time the professional responds. If the student uses an encouraging word or sound, these need to be indicated in parenthesis on the respondent's line.

Numbering the lines of the transcript also helps in the analysis.

Interview with Agent A

1. I: Please tell me what your educational background is?
2. R: Well, let's see . . . I have a bachelor's degree in history and a master's
3. in public
4. administration . . . you know . . . um . . . you don't have to have a master's
5. but it helped
6. me later when I wanted to get promoted . . . (I see). Yeah . . . I noticed
7. my supervisor

8. had gotten an advanced degree.
9. I: So . . . you had your history degree before you were hired . . . you know . . .
10. I mean . . . was the history degree an entry requirement?
11. R: Right . . . well not a history degree specifically, but a college degree. (right)

Students may be tempted to pay a professional transcriber, but if at all possible, transcribing their own notes is another part of the interviewing experience. Hopefully, the students' educational institution has transcribing equipment available to be checked out by students. These machines have a foot pedal to stop and start the tape recorder so the student can keep their hands on the keyboard rather than turning the tape recorder on and off. With digital tape recorders, the students can download the interview and use transcribing software.

Transcribing exact verbiage is critical to analyze how well the students are communicating and also provides important information from the respondent.

Analyzing the Interview

Students will have some sense of how well they think their interview went immediately at its completion. However, it will not be until the transcription is completed that they will have an understanding of how well the objectives of the interview were fulfilled.

Did the interviewee actually answer the questions? Did the interviewee diverge onto a side topic, which might have been interesting, but was not relevant to the objectives of the interview? Was the student able to get the subject back on track?

It is important that the students remember these first interviews are practice. They should be evaluated to learn, not to criticize. The practice interview should be analyzed in a number of areas.

The following areas can be used to help structure the analysis conducted on the practice interview and on future interviews. Students should get in the habit of evaluating each interview conducted so they might learn and improve. The analysis proceeds partially from the transcript and partially from the interviewer's memory. The physical setting and the interviewee's nonverbal cues will be memory driven.

Interview Objectives and Questions

Compare prepared questions with questions actually asked. Were they the same? If not, how did they differ? Why did they differ?

Was the first question sufficiently open-ended to get the interviewee comfortable and talking? Did the questions continue to be open-ended, or did they become more closed? Did the questions motivate the interviewee to answer completely? Were any of the questions leading? Were any of the questions loaded?

Were the interview objectives achieved? If not, why?

Establishing Rapport   Did the interviewer remember to thank the interviewee? Did the interviewer explain the purpose of the interview?

Did the interviewer find some ways to connect with the interviewee? Was the physical setting conducive to increasing the comfort of the interviewee? For example, did it allow the interviewer and interviewee to sit approximately 4 to 6 feet apart?

Was it quiet without any interruptions or distractions? Was there privacy?

Interviewer's Nonverbal Communication   After listening to the interview, how was the rate of speech and the timing of the turn-taking? Did the interviewer ask the questions and include paraphrasing at a pace that was lively, but did not interrupt the narrative and allowed the interviewee to have time to continue if so desired?

How did the interviewer show that he or she was listening? Were encouraging words and expressions used? What about eye contact, facial expression, leaning forward, and open position?

Active Listening   Did the interviewer reflect content and feelings? Did the interviewer use appropriate paraphrasing and summarizing? What factors hindered the ability to listen to the interviewee?

Interviewee's Nonverbal Communication   Were there certain times during the interview that the interviewee's nonverbal cues and verbal response were inconsistent? Was there a pattern in the inconsistency? How did the interviewer react to these changes in behavior?

What were the interviewee's turn-taking cues? Were there any nonverbal cues that indicated certain emotions? What were those emotions, and how were they communicated?

Probing   How many probes were used? What was the purpose for the different probes—to clarify or to elaborate? Did the interviewer interrupt to probe? Did the interviewer wait until the person had finished? Why? Were the interviewee's own words used in the probes? Were there responses that should have been probed further? Were any of the probes leading or loaded?

Summary   What did the interviewer find surprising? What did the interviewer learn from the interview that will help in his or her next interview? What would the interviewer do differently next time? What did the interviewer do best?

Analyzing a Fellow Student's Interview

For additional practice, it is helpful to analyze a fellow student's interview. The student will receive the objectives, original set of questions, the transcript, and ideally, the original tape of the interview.

With exceptions of the areas that needed to be observed, the fellow-student interview analysis will follow the same structure. Time should be allowed to have a verbal exchange between each set of students.

Conclusion

After the students have completed their first interview and analyzed their own and a fellow student's interviews, they will have new appreciation for the difficulties of interviewing. They also should realize they can now consider themselves no longer "rookie" interviewers.

As new interviewers, they still have a great deal to learn, but with practice, these newly acquired skills can continue to be sharpened and polished. With practice comes the need to analyze. Without analysis, practicing the same thing over and over again might just cement bad habits.

As new interviewers, the students are in an optimal position to continue analyzing what works for them and what they should discard. Finding, observing, and learning from effective interviewers in whatever criminal justice career they enter will be one of the most rewarding goals they can achieve.

ALLEN'S WORLD

If it is going to be a professional, formal interview, I agree with Dr. Lord. I always get advance approval for taping. No need to start off on the wrong foot. If there are objections, it gives me the opportunity to discuss them and perhaps overcome the resistance.

I can tell the person the tape is confidential. I can tell the person the tape is being made for his or her protection so there can be no dispute of what was said and the tone, manner, and inflection of what was said. If I am going to transcribe the tape, I can offer a transcript so they will have a record of the questions and responses.

If I think this is going to be a hostile interview, however, I rarely tell the subject in advance that I am going to record. I just show up, put the recorder on the table, and begin asking questions. Most folks are intimidated enough so they do not object. If they do, that to me is an indication of nervousness, that the person has something to hide.

"Why are you reluctant to let me record this interview?" I ask. "What are you afraid of?"

Remember from an earlier chapter, I almost always have the ability to get the attorney I am working for to issue a subpoena.

In North Carolina, one-party consent is needed to tape an interview without telling the other party the conversation is being recorded. For 20 years, I have tried to figure out how this regulation came about.

If I want to tape an interview without the interviewee knowing it, here is what happens. I say to myself, "Allen, do I have your permission to tape this interview?" Allen always says yes. So I can tape the interview and not tell the other person.

I have a small microphone built into the end of a pen. It can record a conversation up to 20 feet away. I put it in my shirt pocket, along with several other pens in case someone asks to use my pen, and I turn on the recorder

Badda bing, badda boom—I have got the interviewee on tape, and he or she does not know it. And, it is legal.

No matter what type of formal interview I am conducting, I always begin the tape with an introduction: my name; my occupation; who I am working for; the date, time, and place of the interview; who I am about to interview; and why. I state the name of the subject, who is likely to listen to the tape, and that it is possible the tape might be played in court.

I always say on the tape who is present during the interview. I get each person to identify themselves, and especially if the tape is going to be transcribed, I get them to identify themselves so the transcriber can recognize the voices and assign the statements with the correct person who made that statement.

Always, always, always, I ask the subject to acknowledge on the tape that the interview is being recorded, and I have authorization to record. At the end of each tape, I make the same closing explanation on the tape as the introduction. And, always, always, always, get the subject to confirm the tape was made with his or her authorization.

Having said that, here is an ethical issue for debate. As I said, I will, at times, use a hidden microphone and miniature recorder to tape an interview without the person's consent.

For example, I had a case in which a man was creating a product in violation of a clause in his hiring contract that said if he left the company, he would not go into competition with that company. Well, he left the company and began duplicating the product of the company he used to work for. I called the man, pretending to be a buyer. He agreed to meet with me at his warehouse and sell me the items I needed.

Common sense tells you that if I walk in with a microphone in hand, he is not going to talk. He is not going to sell me the item. So, I wired myself. Before entering the warehouse, I recorded my name, date, time, address of where I was, and the purpose of my being there. I went in. I made sure I asked him loud and clear, "I'm supposed to meet with John Smith. Are you John Smith?"

I schmoozed a little bit about the warehouse, the equipment, and so forth. We talked about me buying some widgets: "I need ten of them."

John Smith said he would be right back. He disappeared through a door and within seconds came out with a box. In it were 10 widgets identical to the ones being made by my client. John wanted cash. I paid him, telling him I needed a written receipt for my boss so I could get reimbursed. John hand wrote out a receipt, saying the money was for 10 widgets at so much per widget. We shook hands, and I left.

Outside, I finished the tape. I told where I was, the date, time, and that what was on the tape was a secret recording of me buying 10 widgets from John Smith.

Now, if you are brilliant, you are asking yourself, why do you even need to record this transaction? Well, what if we go to court. John Smith denies ever meeting with me. In fact, he has an alibi that he was fishing with friends the day the supposed transaction took place. Without the tape, this becomes a shouting match between his alibi witnesses and me. The tape is conclusive proof that the transaction took place.

No matter what type of interview I have recorded, I always label the tape: time, place, date, and people on the tape. I initial it. If possible, I get the people I interviewed to initial it.

This practice helps with what is called *chain-of-custody*. If this tape is to be introduced at trial, I have to prove it is the original tape. If it has been given to someone to be transcribed, I supply a form for the transcriber. When did she get it? Did she do the transcribing? Is this the transcript? Has the tape, or the transcript, been altered? That form is signed and dated. The chain of custody has been maintained.

Let us hit some of the highlights in Dr. Lord's practice interview:

1. I rarely supply questions in advance. Why is this a good practice for a PI?

2. I do not shy away from areas of controversy. If a policeman I need to interview has a history of complaints, I jump right in. Why do you think?

3. I almost always dress slovenly. I rarely shave. I rarely comb my hair. I wear jeans and T-shirts. Why do I do this?

4. I always pay the person doing my transcription. It ensures accuracy. It ensures completeness. I do not want any attorney accusing me of leaving something out of the interview. Cost is not a consideration. The transcription fee is always included in my invoice and paid for by the client.

5. I do not always tell the subject that I am recording the interview. Discuss the ethical issues.

I would like to close by saying that I love living in Allen's world. When possible, and practical, which is most of the time, I use the same interviewing techniques discussed by my co-author. But there are times when I have to cross that invisible line between following the rules and creating solutions to problems that fall outside the regular parameters. In other words, I do what I find necessary to help and protect my client.

I sometimes pat myself on the back with how brilliant I am. Others look at my tactics and call me unethical. To each his own.

There is a 1987 movie, *Broadcast News,* starring William Hurt and Holly Hunter. The movie is a satirical overview of television journalism and television journalists. In that film, at one point, Hurt and Hunter are discussing ethics. Hunter says something like this to Hurt, "There's a line in the sand you are not supposed to cross." And Hurt responds something like this, "Yeah, but they keep moving it."

My line in the sand is constantly moving.

That is where Allen's world becomes a fun place to frolic.

References

Adams, S. (1996). Statement analysis: What do words really reveal? *FBI Law Enforcement Bulletin*, October, 12–20.

Adams, S. & Jarvis, J. (2006). Indicators of veracity and deception: An analysis of written statements made to police. *Speech, Language and the Law*, 13, 1–22.

Amendola, K., Leaming, M., & Martin, J. (1996). *Analyzing characteristics of police-citizen encounters in high-risk search warrant issuances, domestic disturbances, hostage and barricaded persons incidents, and encounters with fleeing felony suspects.* (Grant 92-IJ-CX-K019). Washington DC: Police Foundation.

American Psychiatric Association. (1994). *Diagnostic and statistic manual of mental disorders*, (4th ed.). Washington, DC: APA.

American Society of Testing Measures. (2005). E2000-05-Standard guide for minimum basic education and training of individuals involved in the detection of deception (PDD). Retrieved from http://www.techstreet.com/cgi on February 13, 2009.

Anderson, D.E., DePaulo, B.M., Ansfield, M.E., Tickle, J.J., & Green, E. (1999). Beliefs about cues to deception: Mindless stereotypes or untapped wisdom. *Journal of Nonverbal Behavior*, 23(1), 67–89.

Ataiyero, K.T. (December 18, 2004). Judge to decide whether N.C. officers went too far to obtain confession from murder suspect. *News and Observer*, A1.

BBC Homepage. Let him have it! The case of Bentley and Craig. Retrieved from http://www.bbc.co.uk/dna/h2g2/A9115229 on November 10, 2009.

Board of Professional and Occupational Regulations (2003). *Study of the utility and validity of voice stress analysis.* Richmond, VA.: Virginia Department of Professional and Occupational Regulations.

Brewer v. Williams. 430 U.S. 387 (1974).

Bull, R. & Milne, B. (2004). Attempts to improve the police interviewing of suspects. In G.D. Lassiter (Ed.). *Interrogations, confessions, and entrapment* (pp. 181–196). Boston: Kluwer Academic/Plenum Publishers.

Bureau of Justice Statistics. (2004). National crime victimization survey, basic screen questionnaire 2004 version. Retrieved from http://www.ojp.usdoj.gov/bjs/cvict.htm#Programs on January 25, 2009.

Burgoon, J.K., Buller, D.B., Ebesu, A.S., & Rockwell, P. (1994). *Communication monographs*, 61, 303–325.

Canter, D. & Alison, L. (1999). *Interviewing and deception.* Brookfield, VT: Ashgate.

Castiglia, C. (1996). *Bound and determined: Captivity, culture-crossing, and white womanhood from Mary Rowlandson to Patty Hearst.* Chicago: University of Chicago Press.

Charlotte-Mecklenburg Police Department. (March 13, 2009). Electronic monitoring of interview/interrogation. In *Interactive Directives Guide*. Retrieved from http://www.charmeck.org/NR/rdonlyres/euyac4abnje433derftkmvqvpx7yr3yyzexowyln4kjakh4l45b2lhxmvjy6yoz3ojgvwwfmtdweb3ocizctchlkeoc/CMPDDirectives.pdf on January 14, 2010.

Charlotte-Mecklenburg Police Department. (2005). Juvenile Operations. In *Interactive Directives Guide*. Retrieved from http://www.charmeck.org/NR/rdonlyres/euyac4abnje433derftkmvqvpx7yr3yyzexowyln4kjakh4l45b2lhxmvjy6yoz3ojgvwwfmtdweb3ocizctchlkeoc/CMPDDirectives.pdf on January 14, 2010.

Collins, R., Lincoln, R., & Frank, M.G. (2002). The effect of rapport in forensic interviewing. *Psychiatry, Psychology, and Law*, 9(1), 69–78.

Colorado v. Spring. 479 U.S. 564 (1987).

Cooper and Cooper v. Chatham County. 455 F.2d 1142; (1972) U.S. App. Lexis 11426.

Crelinsten, R.D. & Szabo, D. (1979). *Hostage-taking*. Lexington, MA: Lexington Books.

Damore, L. (1988). *Senatorial privilege: The Chappaquiddick cover-up*. New York: Bantam Doubleday Dell.

Daubert v. Merrell Dow Pharmaceuticals. Inc. U.S. (1993).

Davis, B. & Mason, P. (2008). Stance-shift analysis: Locating presence and positions in online focus group chat. In S. Kelsey & K. St. Amant (Eds.). *Handbook of Research on Computer-Mediated Communication* (pp. 634–646). Hershey, PA: ICI Press.

DeFabrique, N., Romano, S.J., Vecchi, G.M., & VanHasselt, V.B. (2007). Understanding Stockholm syndrome. *FBI Law Enforcement Bulletin*, 76 (7). Retrieved from http://www.fbi.gov/publications/leb/2007/july2007/july2007leb.htm#page10 on January 14, 2009.

Donohue, W., Ramesh, C., & Borchgrevink, C. (1991). Crisis bargaining: Tracking relational paradox in hostage negotiation. *The International Journal of Conflict Management*, 2 (4), 257–274.

Dunway v. New York. 442 U.S. 200 (1979).

Edelmann, R. (1999). Nonverbal behavior and deception. In D. Canter & L. Alison (Eds.) *Interviewing and deception* (pp. 65–82). Brookfield, VT: Ashgate.

Edwards v. Arizona. 451 U.S. 477 (1981).

Egan, G. (1975). *The skilled helper: A model for systematic helping and interpersonal relating*. Monterey, CA: Brooks/Cole Publishing Co.

Ekman, P. (1996). Why don't we catch liars? *Social Research*, 63(3), 801–817.

Ekman, P., O'Sullivan, M., & Frank, M.G. (1999). A few can catch a liar. *Psychological Science*, 10(3).

Elaad, E. (1999). A comparative study of polygraph tests and other forensic methods. In D. Canter & L. Alison (Eds.). *Interviewing and deception* (pp. 23–40). Brookfield, VT: Ashgate.

Encyclopedia of espionage, intelligence, and security. (n.d.). Retrieved from http://www.espionageinfo.com/ on January 14, 2010.

Erikson, E. (1968). *Identity, youth and crisis*. New York: W.W. Norton.

Espionage Information. (n.d.). Retrieved from www.espionageinfo.com/Pa-Po/Polygraphs on October 2, 2008.

Fatt, J.P.T. (1998). Detecting deception through nonverbal cues: Gender differences. *Equal Opportunities International*, 17, (2), 1–9.

FBI History. (n.d.). Famous cases: Lindbergh kidnapping. Retrieved from http://www.ncapi.org/jm/index.php?option=com_content&view=article&id=44&Itemid=56 on January 14, 2010.

Fisher, R.P. (1995). Interviewing victims and witnesses of crime. *Psychology, Public Policy, and Law*, *1*, (4), 732–764.

Fisher, R.P. & Geiselman, R.E. (1992). *Memory-enhancing techniques for investigative interviewing: The cognitive interview.* Springfield, IL: Thomas.

Frank, M.G. & Ekman, P. (1997). The ability to detect deceit generalizes across different types of high stake lies. *Journal of Personality and Social Psychology*, *72*(6), 1429–1439.

Frazier v. Cupp. 394 U.S. 731 (1969).

Frost, D. (1978). *I gave them a sword: Behind the scenes of the Nixon interviews.* New York: Morrow.

Fuller, C. (2008). High-stakes, real-world deception: An examination of the process of deception and deception detection using linguistic-based cues. Doctoral dissertation at Oklahoma State University.

Fuselier, G. (1986). A practical overview of hostage negotiations. *FBI Law Enforcement Bulletin*, *55*(4), 1–4.

Fuselier, G.D., VanZandt, C.R., & Lanceley, F.J. (1991). Hostage/barricade incidents: High-risk factors and the action criteria. *FBI Law Enforcement Bulletin*, *60*(1), 6–12.

Galianos, J. (n.d.). Brief history of polygraph. Retrieved from http://www.galianospolygraphe.com on October 2, 2008.

Garven, S., Wood, J., Malpass, R.S., & Shaw, J.S. (1998). More than suggestion: The effect of interviewing techniques from the McMartin preschool case. *Journal of Applied Psychology*, *83* (3), 347–359.

Gorden, R. (1998). *Basic interviewing skills.* Prospect Heights, Ill: Waveland Press.

Gudjonsson, G. (1993). *The psychology of interrogations, confessions and testimony.* New York: John Wiley & Sons.

Gudykunst, W.B. (2003). *Cross-cultural and intercultural communication.* Thousand Oaks, CA: Sage.

Hammer, M.R. & Rogan, R.G. (1996). Negotiation models in crisis situation: The value of a communication-based approach. In R.G. Rogan, M.R. Hammer, & C.R. Van Zandt (Eds.) *Dynamic processes of crisis negotiation : Theory, research, and practice.* pp. 9–24, Westport, CT: Praeger.

Hancock, J., Curry, L., & Goorha, S. (2008). On lying and being lied to: A linguistic analysis of deception in computer-mediated communication. *Discourse Processes*, *45*, 1–23.

Hare, R.D. (1978). Psychopathy and electrodermal responses to nonsignal stimulation. *Biological Psychology*, *6*(4), 237–246.

Head, W.B. (1987). The hostage response: An examination of the US law enforcement practices concerning hostage incidents. (Doctoral dissertation. State University of New York at Albany, 1987) *Dissertation Abstracts International 50:4111-A.*

Henriques, D. (December 20, 2008). Madoff scheme kept rippling outward, across borders. *New York Times*, p. A1.

Hess, J.E. (1997). *Interviewing and interrogation for law enforcement.* Cincinnati, OH: Anderson.

Hewlett, M. & Young, W. (September 3, 2009). Abbitt is a free man: Exonerated by DNA after 14 years, he feels a bit lost. *Winston Salem Journal*, p. A1.

Holmberg, U. (2004). *Police interviews with victims and suspects of violent and sexual crimes: Interviewees' experiences and interview outcomes.* Doctoral Dissertation. Stockholm University—Stockholm, Sweden.

Hynan, D. (1999). Interviewing: Forensic psychological interviews with children. *The Forensic Examiner*, March, 25–30.

Inbau, F.E., Reid, J.E., Buckley, J.P., & Jayne, B.C. (2004). *Criminal interrogation and confessions.* (4th ed.) Sudbury, MA: Jones and Bartlett.

Inbau, F.E., Reid, J.E., & Buckley, J.P. (2004). *Essentials of the Reid Technique: Criminal interrogation and confessions.* (4th ed.) Sudbury, MA: Jones and Bartlett.

Jayne, B.C. (1986). The psychological principles of criminal interrogation. In F.E. Inbau, J.E. Reid, & J.P. Buckely (Eds.) *Essentials of the Reid Technique: Criminal interrogation and confessions.* (3rd ed.) (pp. 327–347). Sudbury, MA: Jones and Bartlett.

Jordan, J. (2004). Beyond belief? Police, rape and women's credibility. *Criminal Justice,* 4(1), 29–59.

Kassin, S.M. (1997). The psychology of confession evidence. *American Psychologist,* 52(3), 221–233.

Kassin, S.M. (2005). On the psychology of confessions: Does innocence put innocents at risk? *American Psychologist,* 60(3), 215–228.

Kebbell, M.R. & Wagstaff, G.F. (1999). The effectiveness of the cognitive interview. In D. Canter & L. Alison (Eds.). *Interviewing and deception* (pp. 23–40). Brookfield, VT: Ashgate.

Keyser, J. (2009). Iraqi shoe thrower to be released from jail today. *Associated Press* reprinted in *Charlotte Observer,* September 14, 2009.

Kingston, J. & Stalker, K. (2006). Forensic stylistics in an online world. *International Review of Law Computers & Technology 20,* 95–103.

Kleiner, M. (1999). The psychophysiology of deception and the orienting response. In D. Canter & L. Alison (Eds.). *Interviewing and deception* (pp. 23–40). Brookfield, VT: Ashgate.

Klotter, J.C., Walker, J.T., & Hemmens, C. (2005). *Legal guide for police: Constitutional issues.* (7th ed.). LexisNexis.

Kluckholn, F.R. & Strodtbeck, F.L. (1961). *Variation in value orientations.* Evanston, IL: Row, Peterson.

Knight, D.W. (2000). *Is the polygraph examination a dying investigative tool in the law enforcement community?* Unpublished manuscript.

Koehnken, G. (2004). Statement validity analysis and the "detection of the truth." In P. Granhag & L. Stromwall, (Eds.). *The detection of deception in forensic contexts* (pp. 41–63). Cambridge: Cambridge University Press.

Koppel, M., Schler, J., & Argamon, S. (2009). Computational Methods in Authorship Attribution. *Journal of the American Society for Information Science and Technology* 60, 9–26.

Kupperman, R. & Trent, D. (1970). *Terrorism: Threat, reality, response.* Stanford: Hoover Institution Press.

Leo, R. (2008). *Police interrogation and American justice.* Cambridge: Harvard University Press.

Leo, R.A. (1996). Criminal law: Inside the interrogation room. *The Journal of Criminal Law & Criminology,* 86(2), 266–303.

Lisheron, M. (1978). Mirage. Chicago Sun Times. *American Journalism Review,*

London, K. (2001). Investigative interviews of children: A review of psychological research and implications for police practices. *Police Quarterly,* 4 (1), 123–144.

Lord, V., Davis, B., & Mason, P. (2008). Stance shifts in rapist discourse: Characteristics and taxonomies. *Psychology, Crime & Law* 14, 357–79.

Lord, V. & Gigante, L. (2004). Comparison of strategies used in barricaded situations: SbC and non-sbc subjects. In V. Lord (Ed.). *Suicide by cop: Inducing officers to shoot* (pp. 203–228). Flushing, NY: Looseleaf Law Publications.

Lustig, M.W. & Koester, J. (1999). Intercultural competence: *Interpersonal communication across cultures.* (3rd ed.). New York: Longman.

MacMartin, C. & LeBaron, C.D. (2007). Arguing and thinking errors: Cognitive distortion as a members' category in sex offender group therapy talk. In A. Hepburn & S. Wiggins (Eds.). *Discursive research in practice: New approaches in psychology and interaction.* Cambridge: Cambridge University Press.

Mason, P., Davis, B., & Bosley, D. (2005). Stance analysis: When people talk online. In S. Krishnamurthy (Ed.). *Innovations in e-marketing, II.* (pp. 261–282). Hershey, PA: ICI.

McGinniss, J. (1983). *Fatal vision.* New York: G.P. Putnam's Sons.

McMaster, J.G. (2006). *Civil interviewing and investigating for paralegals.* Upper Saddle River, NJ: Pearson Prentice Hall.

McMains, M.J. & Mullins, W.C. (2006). *Crisis negotiations: Managing critical incidents and hostage situations in law enforcement and corrections,* (3rd ed.). LexisNexis.

Meili, T. (2003). *I am the Central Park jogger: A story of hope and possibility.* New York: Scribner.

Meissner, C.A. & Kassin, S.M. (2004). "You're guilty, so just confess!" Cognitive and behavioral confirmation biases in the interrogation room. In G.D. Lassiter (Ed.). *Interrogations, confessions, and entrapment* (pp. 85–106). Boston: Kluwer Academic/Plenum Publishers.

Michaud, S.G. & Aynesworth, H. (1984). *The only living witness: The true story of serial sex killer Ted Bundy.* Signet Book.

Miranda v. Arizona. 384 U.S. 336 (1966).

New York v. Quarles. 104 S. Ct. 2626 (1984).

Noesner, G. (1999). Negotiation concepts for commanders. *FBI Law Enforcement Bulletin, 68,* 6–14.

North Carolina Association of Private Investigators. (n.d.). Retrieved from http://www.ncapi.org/jm/index.php?option=com_content&view=article&id=44&Itemid=56 on January 14, 2009.

O'Connor, P. (2000). *Speaking of crime: Narratives of prisoners.* Lincoln, NE: University of Nebraska Press.

O'Connor, J. & Seymour, J. (1990). *Introducing neuro-linguistic programming.* Hammersmith, London: Thorsons.

O'Hara, C.E. & O'Hara, G.L. (1994). *Fundamentals of criminal investigation.* (6th ed.). Springfield, IL.: Charles C. Thomas.

Oregon v. Mathiason. 429 U.S. 492 (1977).

Parker, A.D. & Brown, J. (2000). Detection of deception: Statement validity analysis as a means of determining truthfulness or falsity of rape allegations. *Legal and Criminological Psychology, 5,* 237–259.

Peak, K.J. (2009). *Policing American: Challenges and best practices.* Upper Saddle River, NJ: Pearson/Prentice Hall.

Pekerti, A.A. & Thomas, D.C. (2003). Communication in intercultural interaction: An empirical investigation of idiocentric and sociocentric communication styles. *Journal of Cross-Cultural Psychology, 34,* 139–154.

Pezdek, K. & Roe, C. (1997). The suggestibility of children's memory for being touched: Planting, erasing, and changing memories. *Law and Human Behavior, 21,* 95–106.

Pipe, M.E., Orbach, Y., Lamb, M., Abbott, C.B., & Stewart, H. (2006). Do best practice interviews with child abuse victims influence case outcomes? (2006-IJ-CX-0019). Rockville, MD: National Institute of Justice.

People v. Raymond Buckley et al. Los Angeles Sup. Ct. No. A750900 (1990).

Powell, L. & Amsbay, J. (2006). *Interviewing: Situations and contexts.* Boston: Allyn and Bacon.

Pring, R.K. (1990). Logic and values: A description of a new course in criminal justice ethics. In F. Schmalleger (ed.) *Ethics in Criminal Justice* (pp. 164–177). Bristol, IN: Wyndham Hall Press.

Rabon, D. (1992). *Interviewing and interrogation.* Durham, NC: Carolina Academic Press.

Rabon, D. (1994). *Investigative discourse analysis.* Durham, NC: Carolina Academic Press.

Rhode Island v. Innis. 446 U.S. 289 (1980).

Ridgeway, B.J. (1999). The hermeneutical aspects of rapport (Doctoral dissertation) Retrieved from Dissertation Abstracts. (AAT 99377686).

Riessman, C.K. (1993). *Narrative analysis.* London: Sage.

Rudacille, W.C. (1998). *Identifying lies in disguise.* Dubuque, IA: Kendall/Hunt Publishing.

Rutledge, D. (1996). *Criminal interrogation: Law and tactics.* (3rd ed.). Incline Village, NV: Copperhouse.

Samantrai, K. (1996). *Interviewing in health and human services.* Chicago: Nelson-Hall Publisher.

Samovar, L.A. & Porter, R.E. (1991). *Communication between cultures.* Belmont, CA: Wadsworth.

Schmalleger, F. (1990). *Ethics in criminal justice: A justice professional reader.* Bristol, IN: Wyndham Hall Press.

Schroeder, A. (2008). *Presidential debates: Fifty years of high risk TV.* (2nd ed.) New York: Columbia University Press.

Sear, L. & Williamson, T. (1999). British and American interrogation strategies. In D. Canter & L. Alison (Eds.). *Interviewing and deception* (pp. 65–82). Brookfield, VT: Ashgate.

Shearer, R.A. (2005). *Interviewing: Theories, techniques, and practices* (5th ed.). Upper Saddle River, N.J.: Prentice Hall.

Shuy, R. (1998). The *language of confession, interrogation, and deception.* Thousand Oaks, CA: Sage.

Slatkin, A.A. (2002). Intelligence in crisis and hostage negotiations. *Law and Order, 50*(7), 22–28.

State v. Jackson. 308 NC 549, 304 S.E.2d 134 (1983).

State v. Kelekolio. 849 P.2d 58 (1993).

Tully, B. (1999). Statement validation. In D. Canter & L. Alison (Eds.). *Interviewing and deception* (pp. 23–40). Brookfield, VT: Ashgate.

United States of America v. Jeffrey R. MacDonald, No. 75-26-CR-3, 485 F. Supp. 1087; (1979) U.S. Dist. LEXIS 9802.

Vito, G.F., Maahs, J.R., & Holmes, R.M. (2007). *Criminology: theory, research and policy.* (2nd ed.). Sudbury, MA: Jones and Bartlett.

Vrij, A. (2005). Criteria-Based Content Analysis: A qualitative review of the first 37 studies. *Psychology, Public Policy, and Law 22,* 3–41.

Vrij, A., Edward, K., Roberts, K.P., & Bull, R. (2000). Detecting deceit via analysis of verbal and nonverbal behavior. *Journal of Nonverbal Behavior, 24*(4), 239–263.

Walsh, A. (1988). *Understanding, assessing, and counseling the criminal justice client.* Pacific Grove, CA: Brooks/Cole Publishing.

Wetzel, C. (1982). Self-serving biases in attribution: A bayesian analysis. *Journal of Personality and Social Psychology, 43*(2), 197–209.

Whitcomb, C. (2001). *Cold zero: Inside the FBI hostage rescue team.* Boston: Little Brown.

Yeschke, C.L. (1997). *The art of investigative interviewing.* Boston: Butterworth-Heinemann.

Zulawski, D.E. & Wicklander, D.E. (1993). *Practical aspects of interview and interrogation.* Boca Raton, FL: CRC Press.

Index

abbreviation, 73
abuse, 20
acceptance, 12, 181
accuracy
    of children, 161
    of meaning, 8
    paraphrasing, 113
    of polygraph, 264
accusation, 124
active listening, 290
actualization, 11
acute stress disorder (ASD), 148
ADA (assistant district
        attorney), 190
adolescents, 103, 157
adultery, 196
adults, 147–151, 164
advice, 30, 254, 270
affidavit, 123, 143
age, 152, 222
agency, 279
agreement, 170, 181
alcohol, 10
alibi, 48
    friends giving, 271
    impeachment of, 224
    suspect, 222
aliens, 177
alimony, 193
allegation, 130, 155
Ames, Aldrich, 266–267
analysis, 132–134; See also
        psychoanalysis; stance-
        shift analysis
    criterion-based, 133
    DNA, 206

evidence, 26, 227
interview, 289–290
meta, 81
quick, 262
semantic, 135–137
statement, 274–276
technique of, 281
test-data, 259
voice-stress, 187
androgyny, 173
answers
    abbreviated, 73
    clarification to, 58
    focus on, 278
    information vs., 206
    relevant, 35
antecedents, 212
antipathy, 36
anxiety, 267
    lying and, 223
    reduction of, 12–14
appointments, 28
    making, 57
    office, 174
argument, 170
Arizona v. Roberson, 219
articulation, 180
ASD (acute stress disorder), 148
Asians, 178
assault, 28, 45, 147
assertiveness, 23, 112
assessment, 105, 117
assistant district attorney (ADA), 190
assumption, 113, 163
ATF (Bureau of Alcohol, Tobacco,
        and Firearms), 252

atmosphere, 108
attitude, 114
attorney, 24, 270; See
        also defense
    court, 103
    criminal defense, 190–193,
        202–206
    interview tips for defense,
        189–190
    provided, 25
attunement, 250
audiotape, 123, 289
authority
    figures of, 152
    legitimate, 207
    limited, 241
    polite, 66
    positions of, 10
    symbolic, 251
authorization, 292

background, 47, 235
backward recount, 48
badgering, 166
balance, 133
bargaining, 249
barricades, 239
barriers
    cultural, 167
    to intervention, 69
    psychological, 10
bartenders, 207
baseline, 181
bedwetting, 152
behavior
    adapter, 214

behavior (*continued*)
  attending, 74–75
  baseline, 215
  deprecating, 182
  disturbances of, 152
  during eye contact, 80
  immoral, 18
  nonverbal, 54
  perspective on, 114
belief
  feigned, 13
  system of, 5
belonging, 11
Bentley, Derek, 5
bias, 14
  interviewer, 49
  opinion, 105
  reducing, 281
blame, 54
  deflecting, 60
  shifting of, to event, 280
blinking, 214
blood, 225
bluntness, 183
body
  activity, 80
  language, 119
  posture, 100
  space, 176–177
bond
  posting, 204
bowing, 177
brainstorming, 248
Branch Davidian-Waco
      incident, 253
breathing, 72, 259
brevity, 135
Brewer v. Williams, 217–218
Bureau of Alcohol, Tobacco,
      and Firearms
      (ATF), 252
bureaucracy, 91
bystanders, 243

caseworker, 117
Caucasians, 178
caution, 78
CC (court counselor), 106

chain-of-custody, 130, 293
characteristics
  cultural, 182–183
  negotiator, 240–241
charges, 33
  dismissed, 60
  felony, 164
children, 55, 246
  accuracy of, 161
  care of, 171
  cognitive factors of, 153–155
  development of, 152–153
  embellishment by, 166
  interview of, 151
  minor, 257
  older, 157
  parents *vs.*, 276
  preschool, 154
  victimized, 147
  young, 156
choice, 253
chronology, 131
civil court, 193
clarification
  answer, 58
  comment, 68
  information, 158
  message, 7
class, 182
client
  defense, 18
  international, 167
  middle-class, 92
  potential, 32
  rambling, 121
  resistant, 114–116
  treatment of, 112
coaching, 106
coercion, 131
collectivism, 169–170
Colorado v. Spring, 219
comfort, 78
comment, 68
communication
  basis of, 3–5
  deception related to, 78
  foundation of, 16
  indirect, 169

inhibitors of, 9–11
intercultural, 179–181
nonverbal, 63, 76–84,
      175–176, 290
as process, 5–9
therapeutic, 247
two-way, 183
willingness to, 80
written, 85
community
  conscience, 219
  resources, 150
compensation, 197
competency, 16
competitors, 169
complainants, 264
complaints, 185
compliments, 67
comprehension, 4
computer voice stress analyzer
      (CVSA), 267–268
computers, 259, 270
concentration, 149
condescension, 184
conditions, 36
conduct, 98
confession, 22
  acceptance of, by courts,
      219–221
  false, 233–234
  of guilt, 127
  internalizing, 210
  involuntary, 217
  psychology of, 211–213
  Supreme Court on, 219
  validity of, evidence, 22–25
confidence, 15
  suspect's, 226
  vocal, 89
confidentiality, 16, 26–27
  parameters of, 91
  promising, 193
confrontation, 115
confusion, 7
conscience, 30
  absence of, 266
  community, 219
consciousness, 148

consent, 217
consequence, 25, 212
Constitution, U.S., 130, 216
contact information, 232
content
    reflecting on, 98–99
    verbal, 80
context, 69
    event, 164
    high *vs.* low, 170–171
conversation, 17, 108
    peaceful, 252
    starting, 202
conviction, 135
Cooper and Cooper v. Chatham
    County, 222
cooperation, 41
coping
    resources, 148
    skills, 241
corrections, 142
corroboration, 203
corruption, 33
counseling
    long-term, 112
    PI giving, 254
countermeasures, 265–266
counterterrorism, 263–264
court; *See also* civil court
    confession denied by, 221
    identification used in, 142
court counselor (CC), 106
Craig, Chris, 5
creativity, 92
credit cards, 205
crime
    future, 148
    perception of, 37–38
    personal, 147
    prevention of, 96
criminal defense, 202–206
crisis
    intervention, 238–239, 254
    management, 246
    negotiation, 237–238
    situation dynamics, 241–242
criteria measurements, 252–253
criticism, 183

cross-check, 105
cross-examination, 193
cues
    leakage, 78
    paralingual, 83–84
    visual, 76
culpability, 211
culture
    Asian, 178
    barriers to, 167
    communication between,
        179–181
    differences within, 167–168
    generalization of, 181
    global characteristics of,
        182–183
    horizontal *vs.* vertical, 171–172
    taboos of, 176
curiosity, 14, 51
custody, 217
CVSA (computer voice stress
    analyzer), 267–268

database, 270
Daubert v. Merrell Dow
    Pharmaceuticals, Inc., 265
Davis, Boyd, 132
death, 250
debate, 175
    ethics, 292
    polarized, 268
deception, 34
    communication related to, 78
    examinee, 265
    grades of, 260
    motivation of, 213–215
    permitted, 211
    phrases of, 274
    stress in act of, 268
    victim's, 215
defendant, 162
defense; *See also* attorney
    client, 18
    criminal, 190–193, 202–206
    presentation of, 201
    reasonable doubt created
        by, 190
    team, 143

defensiveness, 42
delivery, 72
delusions, 243
demeanor, 90
denial, 13, 229
department of motor vehicles
    (DMV), 196
depersonalization, 136
depression, 240
detail, 45
    created, 24
    fabrication of, 137
    factual, 126
    location, 203
detectives, 206
detention, 217
determination, 16, 115
development
    of children, 152–153
    of language, 155–156
    objective, 95
    question, 38
diagrams, 199
dialogue, 47, 77
dignity, 17
disabilities, 65
disclosure, 143
discovery, 162
discrepancies, 80
dismissals, 192
disorders, 240
disruptions, 88, 121
dissonance, 223
distortion, 44
diversity training, 172
divorce, 255
DMV (department of motor
    vehicles), 196
DNA analysis, 22, 206
documentation
    fraud, 198
    interview, 123–124
dolls, 156
domestic cases, 193–197, 254
dominance, 149, 227, 243
dress
    casual, 92
    interview, 287

drugs, 10, 104
dry mouth, 79
due-process, 20
Dunway v. New York, 216
dynamics
    crisis situation, 241–242
    hostage situation, 242–245
    targeting, 39

education, 40
Edwards v. Arizona, 218
ego
    definition of, 213
    threats to, 42, 104, 159
Ekman, Paul, 81
elaboration, 98
    interviewee, 280
    promoting, 111
    story, 224
embarrassment, 130
embellishment, 166
embezzlers, 31
emblems, 81
emergency medical technician
        (EMT), 199
emotions
    defusing, 49
    involvement of, 70
    regression of, 152
    states of, 128
    strong, 150
    venting, 66
    words bring, 275
empathy, 97
    cultivating, 56
    deep, 119
    maximizing, 42–43
    relaying, 160
Employee Polygraph Protection
        Act (EPPA), 261
employment, 40
employment screening, 260
EMT (emergency medical
        technician), 199
encouragement
    indication of, 108
    recollection, 43
    sounds of, 288

enticement, 160
epilogue, 133
EPPA (Employee Polygraph
        Protection Act), 261
equipment, 124
Erikson, Erik, 152
errors, 129
Escobedo v. Illinois, 218
esteem, 11
ethics
    debate about, 292
    issues of, 28–29
    rules of, 31
    standards of, 20–22
    surrounding interview, 19–20
    tools pertaining to, 25–26
evaluation, 104–107, 241
evasiveness, 274
events
    blame on, 280
    context of, 164
    family, 174
evidence
    admissible, 195
    adultery, 196
    analysis, 26, 227
    collected, 65
    false, 220
    immaterial, 140–141
    physical, 138
    preponderance of, 216–217
    review of, 263
    soliciting, 209
    statement *vs.*, 225
    validity of confession, 22–25
examination, 206
examinee, 265
exceptions, 59
expenses, 29, 194
experience, 147
    interview, 46
    interviewer, 107
    knowledge gained
        from, 70
    offender's, 230
    threatening, 101
    variety of, 4
explanation, 128

expression
    facial, 76, 177–178
    feelings needing, 16
    message, 81
eye contact, 19
    behavior during, 80
    direct, 178–179
    good, 75
    welcoming, 90

fabrication, 44, 137
face, 25
    expression of, 76, 177–178
    saving, 250
facts, 21, 126
fame, 22
family events, 174
fantasy, 153
fatigue, 21, 74
feedback, 4
    negative, 79
    peer, 56
    providing, 71
feelings, 101, 253
    concrete, 112
    expression of, 16
    handling, 117
    probationer, 39
    reflection of, 100–104, 151
    respect, 68
fees, 29, 32
felonies, 164
female, 182, 245
femininity, 172–173
Fifth Amendment, 216–218, 234
filtering, 6, 74
Fisher, R. P., 44–45
flattery, 183
flippancy, 52
flow, 47, 108, 200
forensics, 158
formality, 123
format, 128–129
forms, 169–170
formula, 285
Fourteenth Amendment, 219, 234
Fourth Amendment, 216, 234
Fraizer v. Cupp, 219

frame, 112
fraud, 198
freedom, 51
friends, 255, 271, 290
frowning, 115

gender, 6
Gender Genie, 275
generalization, 104, 181
gesture, 12, 81–83, 119
girls, 166
goals, 26, 116
Gorden, Raymond, 41–42, 90–91
government, 27, 202
grammar, 130
grooming, 83
guilt, 102
    confession of, 127
    pleading, 210

habeas corpus, 218
handicaps, 222
handshake, 63, 93
handwriting, 82, 127
Hanssen, Robert Philip, 267
Hare, Robert, 266
hearing, 9, 73–74, 143
hearsay, 120
hesitation, 54, 100, 151
hierarchy-of-needs, 11
holidays, 174
home visits, 89
hostage taker
    female, 245
    motivation, 238
hostages, 242–245
hostility, 83, 291
humor, 180
hypothesis, 44

IA (Internal Affairs), 260
id, 213
IDA (investigative discourse analysis), 132–134
ideas, 58
identification
    court use of, 142

eyewitness, 271
positive, 195, 245
of professional, 286
proper, 191
by woman, 163
idiom, 40
images, 3
imagination, 226
impact, 77
impairment, 65
impatience, 46
impeachment
    alibi, 224
    grounds for, 203
    witness, 163
importance, 73
impressions, 66, 100
improvisation, 43, 150
impulse, 228
incentive, 17, 154
incrimination, 233
individualism, 169–170
inference, 273
informants, 228, 263
information, 21
    answers *vs.*, 206
    clarifying, 158
    collecting, 84
    common, 168
    contact, 110, 232
    evaluation, 104–107
    extraneous, 49
    fabricated, 106
    planting, 235
    recall, 153
    summarizing, 67
    supplying, 58
    valid, 55
injury, 197–202
innocence, 31, 166
innuendo, 119
instinct, 187
insurance, 197
integrity, 98, 192
intelligence
    counter, 261
    low, 79
intention, 30

interest
    genuine, 43
    projecting, 75
Internal Affairs (IA), 260
Internet, 200, 265, 287
interpretation, 4, 119
    care in, 136
    filtering before, 74
    limits of, 78
interrogation, 22, 263; *See also* Reid Nine Steps of Interrogation
    approaches to, 224–228
    end of, 232–233
    interview *vs.*, 209–210
    legal framework of, 216–222
    postindictment, 218
    strategies of, 212–224
    techniques, 219–221, 230–231
interrogator, 216
interview
    affidavit generated from, 143
    analysis, 289–290
    areas, 88–89
    beginning of, 63–64
    booking, 36
    of children, 151
    coaching, 106
    conducting, 157–158, 287–288
    defendant's, 162
    defense attorneys and, 189–190
    documentation, 123–124
    dress, 287
    employee, 199
    ethics surrounding, 19–20
    follow-up, 149
    forensic, 158
    foundations of, 3
    helping, 64, 68–69, 111–114, 141
    high-stakes, 281
    inconsistencies within, 67
    interrogation *vs.*, 209–210
    investigative, 141
    language of, 155–156

interview (*continued*)
  objectives for, 36–38
  in office setting, 93
  phases of, 64–69
  police, 159
  practice, 285–286
  preparing to, 26, 35
  pretest, 257
  principles, 14–16
  scheduling, 258, 286
  short, 125
  student, 290–291
  techniques, 192
  tenor of, 59
  termination of, 161
  transcription, 288–289
  of victims, 147–151
interviewee, 290
  comfort, 78
  conviction, 135
  disabilities, 65
  elaboration, 280
  hostility, 291
  present tense for, 118
  questions asked by, 161
  recapitulation, 109
  review, 129
  steering of, 39
  vocabulary of, 40
interviewer, 290
  bias, 49
  complaints, 185
  experience, 46, 107
  fatigue, 21
  flexibility of, 14
  introduction by, 180
  manner, 53
  method, 19
  police-affiliated, 237
  preparation, 26
  psychologist, 165
  reinforcement, 154
  report made by, 131
  rookie, 291
  self-confidence of, 15
intimacy, 176
introduction, 64
  interviewer's, 180

plan, 90
investigation, 59, 201
investigative discourse analysis (IDA), 132–134
investigative interview, 64, 141

jail, 248
journalism, 57
judge, 143
judgment, 14, 54
jurisdiction, 165
jury, 22
justice, 3
justification, 13, 19
Justification Theme Development, 210–211
juveniles, 103

Kassin, S. M., 22–25
kinesthetics, 84

language
  body, 119
  common, 168
  development of, 155–156
  street, 40
Lanier v. South Carolina, 216
lateral eye movement (LEM), 86–87
Latin Americans, 172
law enforcement of, 79
leakage hierarchy, 80
legality, 216–222
  complex, 142
  Supreme Court and, 190
  terms of, 156
legitimacy, 157
Lego v. Twomey, 216
LEM (lateral eye movement), 86–87
lexicon, 40
license plate, 194
licensing, 189, 194
lie detector, 186; *See also* polygraph
lifestyle, 190
lineup, 191
linguists, 132

listening
  active, 64, 99, 290
  hearing differentiated from, 9, 73–74
literacy, 131
literature, 268
location, 203
logic, 170
love, 11
loyalties, 97
lying, 32, 226, 255

MacDonald, Jeffery, 138–141, 224–226
males, 243
management, 246
mandate, 115
manhood, 254
manipulation, 273
masculinity, 172–173
Maslow, Abraham, 11
Mason, Peyton, 132
maturity, 147
mean length of utterance (MLU), 134–135
meaning
  accuracy of, 8
  unintended, 6
media, 244
medication, 46, 240
meetings, 114
memory
  questions enhancing retrieval of, 43–48
  repression of, 155
  stimulating, 48
message, 7
method, 19
Mexicans, 178
microphones, 93
Middle Easterners, 172–173
Miles, Sidney, 5
mimicry, 72
minors, 257
Miranda rights, 24, 220, 234
Miranda warning, 25, 129, 221
mirroring, 72–73
misappropriations, 58

misconduct, 264
misdemeanors, 164
misrepresentation, 219
misunderstanding, 5
MLU (mean length of utterance), 134–135
motivation
    anticipatory, 91
    for deception, 213–215
    hostage taker, 238
    human needs and, 11
    questions encouraging, 40–42
    of witness, 95–97
motivational impairment effect, 83–84
mouth, 81
MRI, 269
multitasking, 173

narrative, 204
National Institute of Child Health and Human Development (NICHD), 158
negligence, 197
negotiation, 115
    approaches to, 249–252
    crisis, 237–239
negotiator
    characteristics, 240–241
    media used by, 244
    property damage by, 251
    subject relating to, 250
neighbors, 207
Neurolinguistic programming (NLP), 84
neurolinguistics, 73, 84–88
neutralization, 229
New York v. Quarles, 221
newspaper, 255
NICHD (National Institute of Child Health and Human Development), 158
NLP (Neurolinguistic programming), 84
nodding, 108, 175
noise, 6
nuance, 119

objectives, 35–38, 95, 287
observations, 46
    reaction, 53
    third member, 71, 126
offender, 239
offenses, 3
office
    appointments, 174
    pictures in, 186
Office of Professional Responsibility (OPR), 260
omission, 44
openness, 20
opinion, 105
OPR (Office of Professional Responsibility), 260
Oregon v. Mathiason, 217
orientation, 5, 168

PACE (Police and Criminal Evidence Act), 210
pacing, 118
paralinguistics, 76, 83–84
parameters, 293
paranoia, 243
paranoid schizophrenia, 240
paraphrasing, 98, 113
parents, 164, 276
parroting, 99
partners, 173
passivity, 242
patience, 56
patronizing, 184
pause, 125
pay, 173, 196
peers, 155, 228
people
    ranking, 171
    street, 207
People v. Buckey, 154
perception
    of crime, 37–38
    of culpability, 211
    of professionals, 172
    of time, 173–174
performance, 8
permission, 291

perpetrator, 227
persistence, 97
personal injury, 197–202
personality, 19
perspective, 114
perspiration, 259
persuasion, 8
petition, 37
phone numbers, 200
phrase
    deceptive, 274
    slang, 102
    tentative, 118
PI (private investigator)
    code for, 29
    criminal defense attorneys used by, 190–193
    giving counsel, 254
    licensing, 194
    preparation of, 56
    privacy rights of, 92
    restrictions of, 30, 234
    tenets, 189
    working with witness, 142
pictures taking, 195
pitch, 83, 214
plaintiff, 198
play, 157
ploys, 169
polarization, 9
police, 20
    detectives, 206
    interview, 159
    interviewer affiliated with, 237
    reports, 162, 199
Police and Criminal Evidence Act (PACE), 210
policy, 51, 54
polygraph, 23
    beating, 265–267
    community, 266
    machines, 187
    purpose of, 260–264
    reliability of, 264–265
    Supreme Court on, 265
    training, 262
    use of, 257–259
positivity, 67

posture, 72
  body, 100
  open, 75
  sinking, 257
power, 27, 246
practice, 119, 285–286
preadolescents, 157
precautionary measures, 210–211
precontact, 65
prediction, 149, 161
prejudice, 14
preparation, 161
  interview, 26, 35
  PI, 56
  victim, 91
pressure, 228
pretense, 33
pride, 15
Pring, Robert, 25–26
priority, 116
prison, 41
privacy, 92, 197
private investigator. See PI
privilege, 37
probation, 41
probationer, 36, 39
probing, 107–111, 142, 290
  effective, 95
  follow-up, 35
  loaded, 109
  notes, 124–126
  strategy, 47
process
  communication as, 5–9
  of stress-reaction, 148
professionalism, 70, 198
professionals, 172
proffer, 143
projection, 14
  of interest, 75
  of suspicion, 52
prologue, 133
prompts, 158
pronouns, 275
props, 88
protest, 231
psychoanalysis, 213
psychologists, 165

psychology, 152, 169
  barriers in, 10
  of confessions, 211–213
  discursive, 132
  evaluations based on, 241
public places, 195
pulse, 140
pupil dilation, 214
puppets, 155
Pyles v. State, 217

questions, 289
  in advance, 293
  asking, 86
  basic, 38–40
  closed-ended, 49–50
  control, 53–54
  corresponding, 287
  formulating, 159–161
  future-oriented, 234
  hesitation absent from, 100,
    151
  hypothetical, 52–53
  implication, 230–231
  improper, 216
  interviewee asking, 161
  leading, 55
  memory retrieval enhancing,
    43–48
  motivating, 40–42
  neutral, 159
  nonthreatening, 51–52
  open-ended, 38, 48–49, 200
  opening, 117
  pacing of, 118
  predetermined, 259
  privacy and, 197
  sequencing types of, 50
  social-incentive, 154
  trap, 231
  why, 54–55

ranking, 171
rape, 206
rapport
  building, 47, 71, 157
  developing, 154, 180
  establishing, 63, 69–72, 290

rationalizations, 13, 211, 223
ratting out, 51
reaction, 79–80
  observing, 53–54
  stress, 148
reasonable doubt, 190
recall, 45, 153
recapitulation, 109
receipts, 205
reciprocal-discovery, 162
recognition, 22
recollection, 43
recommendations, 30
recording, 124
  audio, 66
  authorization, 292
  tape, 31
records, 198
referrals, 150
regression, 152
rehabilitation, 229
Reid, J. E., 228
Reid Nine Steps of
    Interrogation, 228–230
relationship building, 64
reluctance, 96
repetition, 183
reporting, 29
reports, 131, 162, 199
representative, 27, 202
reputation, 191
requests, 87
research, 27
resonance, 86
resources
  community, 150
  coping, 148
respect, 12, 68
respiratory patterns, 85
response, 278
  canned, 287
  completeness of, 125
  general, 52
  generating, 95
  incriminating, 233
  length of, 99
  restricted, 111
restitution, 3

restrictions, 30, 234
retardation, 222
retention, 44, 193
review
    evidence, 263
    interviewee, 129
Rhode Island v. Innis, 210, 218
rhythm, 72, 121
rights, 24, 130
roles, 69, 111
Rule 707, 265
rules
    ethics, 31
    following, 293
Rutledge, Devilish, 209

safety, 11
scheduling, 258
schizophrenia, 240
screening, 260–261
seating, 186
security, 11
self-confidence, of interviewer,
    15
semantics, 74, 135–137
sender, 167
senses
    dominant, 84
    shifting, 86
Sept. 11th 2001, 262
setting, 45, 88–91, 201
sex
    assault, 45, 147
    orientation of, 168
shame, 103
shorthand, 121
signature, 82, 129
silence, 24
    active, 108
    use of, 77
SIRR (suggestive questions,
    social influence,
    reinforcement, and
    removal from direct
    experience), 155
Sixth Amendment, 218–219, 234
skills
    coping, 241

critical, 73
evasive, 107
slang, 40, 102
smiling, 214
social factors, 156
society, 154, 182
software, 289
sounds, 288
sources, 21, 193
South Americans, 172
southerners, 185
space, 176–177
speaking, 180
special needs, 161
Specialized Weapons and
        Tactics (SWAT), 237
speech
    figures of, 72
    parts of, 277
    rate of, 77
    shifts in, 273
stalking, 195
stalling, 254
stance-shift analysis, 132, 273,
        276–281
staring, 151
State v. Jackson, 220, 228
State v. Kelekolio, 219
statement(s)
    analysis, 274–276
    challenges to, 129–130
    completing, 141
    contextual, 159
    evidence *vs.*, 225
    format of, 127–129
    introductory, 223–224
    matching, 165
    purpose of, 126–127
    vague, 113
status, 177
stenographer, 127
Stockholm Syndrome, 245–246
story
    collapse of, 60
    elaboration, 224
strategy, 47, 222–224
stress
    analysis, 187

deception carrying, 268
reaction process, 148
reactions to, 79–80
reducing, 83
students, 286, 290–291
style, 59
subjects, 215
    disarming, 120
    negotiator relating to, 250
    types of, 239–240
subpoena, 93, 291
suggestibility, 23
    high, 156
    testing, 153
suggestion, 217
suggestive questions, social
        influence,
        reinforcement, and
        removal from direct
        experience (SIRR), 155
summery, 116, 179
superego, 213
superficiality, 101
Supreme Court
    on confessions, 219
    Constitution interpreted by,
        216
    equality of, 173
    legality and, 190
    on Miranda rights, 24
    on polygraph, 265
surrender, 239
surveillance
    domestic, 254
    lunchtime, 194
    setting-up, 196
suspects
    alibi, 222
    confidence, 226
    expectancy, 234
    potential, 209
    provoking, 210
    reluctant, 51
    talking to, 48
    youthful, 221–222
suspicion, 52
SWAT (Specialized Weapons
        and Tactics), 237

symbols
  abstract, 3
  authority, 251
sympathy
  badgering eliciting, 166
  trapped in, 42
syntax, 130

taboo, 176
tactics, 54
talk
  candid, 88
  reluctance to, 96
  suspect, 48
tapes
  challenged, 143
  digital, 289
teachers, 172
technique(s)
  analysis, 281
  automated, 274–275
  coercive, 21
  interrogation, 219–221
  interviewing, 192
  justification of, 19
  neutralization, 229
  rapport building, 71
teens, 164
telephone, 88
tense, 137
tension, 244
terrorism, 263–264
testing
  drug, 104
  suggestibility, 153
threats, 215
  to ego, 42, 104, 159
  to manhood, 254
  questions mitigating, 51
time, 118
  keeping, 288
  night, 204
  perception of, 173–174
  stalling for, 254
tools, 25–26, 159, 246

topic control, 107–108
touch, 176
  appropriate, 153
  social, 182
training, 194
  assertiveness, 112
  diversity, 172
  polygraph, 262
transcription, 142, 288–289
translation, 168
trapped offenders, 239
trauma, 43
  bonding, 246
  defining, 147
treatment, 45, 112
truancy, 39
trust
  instinct, 187
  minimal, 64
trustworthiness, 70
truth
  complete, 224
  creative, 82
  verification of, 258

understanding, 10, 168
*United States of America v.
    Jeffrey R. MacDonald,*
    138, 226
United States v. Garibay, 222
United States v. Scheffer, 265
usage pattern, 275
utterance, 134–135

validity, 55, 104
values
  consequence *vs.*, 212
  gender, 6
verification, 21, 258
victims, 41
  deception by, 215
  interview of, 147–151
  passivity, 242
  preparing, 91
videotape, 123

digital, 289
  taking, 195
vigilance, 148
violence
  of content, 252
  domestic, 103
visitation, 131
vocabulary, 40
voice, 84
  analysis, 187
  attributes, 142
  confidence in, 89
volume, 72
voluntariness, 124

Waco incident, 253
witness
  becoming, 191
  eye, 24
  handshake, 93
  identification by, 271
  impeachment, 163
  key, 29
  list, 198
  motivation of, 95–97
  PI working with, 142
woman
  identification by, 163
  southern, 185
words
  communication with, 85
  concrete, 167
  emotional, 275
  feeling, 101
  loaded, 55
  manipulating, 273
  short, 109
  stance, 277
  tentative, 120
workmen's compensation, 197
work-product doctrine, 143

Yeschke, Charles, 11
youth, 221–222